Advanced Thin Films and 2D Materials: Mechanism, Fabrication, Characterization and Application

Advanced Thin Films and 2D Materials: Mechanism, Fabrication, Characterization and Application

Guest Editors

Sebastiano Vasi
Ulderico Wanderlingh

MDPI

Basel • Beijing • Wuhan • Barcelona • Belgrade • Novi Sad • Cluj • Manchester

Guest Editors
Sebastiano Vasi
Department of Mathematics
and Computer Sciences,
Physical Sciences and Earth
Sciences (MIFT)
University of Messina
Messina
Italy

Ulderico Wanderlingh
Department of Mathematics
and Computer Sciences,
Physical Sciences and Earth
Sciences (MIFT)
University of Messina
Messina
Italy

Editorial Office
MDPI AG
Grosspeteranlage 5
4052 Basel, Switzerland

This is a reprint of the Special Issue, published open access by the journal *Materials* (ISSN 1996-1944), freely accessible at: https://www.mdpi.com/journal/materials/special_issues/QG3I7624T6.

For citation purposes, cite each article independently as indicated on the article page online and as indicated below:

Lastname, A.A.; Lastname, B.B. Article Title. *Journal Name* **Year**, *Volume Number*, Page Range.

ISBN 978-3-7258-3887-5 (Hbk)
ISBN 978-3-7258-3888-2 (PDF)
https://doi.org/10.3390/books978-3-7258-3888-2

Contents

materials

MDPI

Review

Optical Second Harmonic Generation on $LaAlO_3/SrTiO_3$ Interfaces: A Review

Andrea Rubano [1,2,*] and Domenico Paparo [2]

1 Physics Department "E. Pancini", University Federico II, Monte S. Angelo, Via Cintia, 80126 Naples, Italy
2 Institute of Applied Sciences and Intelligent Systems (ISASI), Consiglio Nazionale delle Ricerche (CNR), Via Campi Flegrei 34, 80078 Pozzuoli, Italy; domenico.paparo@cnr.it
* Correspondence: andrea.rubano@unina.it

Abstract: As we approach the limits of semiconductor technology, the development of new materials and technologies for the new era in electronics is compelling. Among others, perovskite oxide hetero-structures are anticipated to be the best candidates. As in the case of semiconductors, the interface between two given materials can have, and often has, very different properties, compared to the corresponding bulk compounds. Perovskite oxides show spectacular interfacial properties due to the the rearrangement of charges, spins, orbitals and the lattice structure itself, at the interface. Lanthanum aluminate and Strontium titanate hetero-structures ($LaAlO_3/SrTiO_3$) can be regarded as a prototype of this wider class of interfaces. Both bulk compounds are plain and (relatively) simple wide-bandgap insulators. Despite this, a conductive two-dimensional electron gas (2DEG) is formed right at the interface when a $LaAlO_3$ thickness of $n \geq 4$ unit cells is deposited on a $SrTiO_3$ substrate. The 2DEG is quite thin, being confined in only one or at least very few mono-layers at the interface, on the $SrTiO_3$ side. A very intense and long-lasting study was triggered by this surprising discovery. Many questions regarding the origin and characteristics of the two-dimensional electron gas have been (partially) addressed, others are still open. In particular, this includes the interfacial electronic band structure, the transverse plane spatial homogeneity of the samples and the ultrafast dynamics of the confined carriers. Among a very long list of experimental techniques which have been exploited to study these types of interfaces (ARPES, XPS, AFM, PFM, ...and many others), optical Second Harmonic Generation (SHG) was found to be suitable for investigating these types of buried interfaces, thanks to its extreme and selective interface-only sensitivity. The SHG technique has made its contribution to the research in this field in a variety of different and important aspects. In this work we will give a bird's eye view of the currently available research on this topic and try to sketch out its future perspectives.

Keywords: non-linear optics; perovskite hetero-structures; 2D electron gas

Citation: Rubano, A.; Paparo, D. Optical Second Harmonic Generation on $LaAlO_3/SrTiO_3$ Interfaces: A Review. *Materials* **2023**, *16*, 4337. https://doi.org/10.3390/ma16124337

Academic Editors: Sebastiano Vasi and Ulderico Wanderlingh

Received: 20 April 2023
Revised: 30 May 2023
Accepted: 31 May 2023
Published: 12 June 2023

1. Introduction

Nowadays, almost every modern electronic device, such as computers, music players, mobile phones, as well as opto-electronic devices [1], is equipped with chip-sets made of semiconductors. However, the bulk characteristics of most semiconducting materials are rather simple. Their true advantage and their functionality lie within the transport of charge carriers across or along the interfaces between different materials. Prominent examples are p–n junctions, Schottky contacts and diodes, as well as various transistor designs [2]. Around the year 2000, the progress in the field of semiconductor technology was described by Moore's law, which holds that the number of transistors that can be placed on an inexpensive integrated circuit doubles approximately every two years. For years, the validity of Moore's law was supported by refined chip-set fabrications processes that allowed for rapidly shrinking transistor sizes. However, around the year 2010, the trend started to break with indications that the doubling in transistor capabilities would occur every three years instead. Approaching the limits of semiconductor technology, the search for new materials and new interface phenomena is inevitable. A promising candidate are so-called complex oxide hetero-structures. This material class was found to show spectacular interfacial properties that are not present in the majority of the constituents due to the rearrangement of charges, spins, orbitals and the re-balancing of the lattice structure at the interface. Notorious examples are $BaTiO_3/SrTiO_3/CaTiO_3$ hetero-structures that show an enhancement of polarization compared to pure $BaTiO_3$, caused by the breaking of the inversion symmetry in the hetero-structure [3]; non-superconducting $BaCuO_2$ and $SrCuO_2$ that show superconductivity when grown as a hetero-structure [4]; and antiferromagnetic $CaMnO_3$ and paramagnetic $CaRuO_3$, which show ferromagnetism at the interface [5].

In 2004, A. Ohtomo and H.Y. Hwang at Bell Laboratories discovered that, at the interface between $LaAlO_3$ and $SrTiO_3$, two textbook band insulators, a conducting two-dimensional electron gas (2DEG) emerges depending on the thickness of $LaAlO_3$ and the precise atomic stacking sequence [6]. This discovery unleashed an intense research effort in order to understand this unexpected new property that possesses huge potential for future applications [7–14]. Most recently various new phenomena in $LaAlO_3/SrTiO_3$ interfaces were reported, ranging from superconductivity [15] and ferromagnetism [16] to tip-induced conductance [17]. In particular, the coexistence of both superconductivity and ferromagnetism, which are normally two mutually exclusive phenomena as ferromagnetism destroys the singlet correlations responsible for the pairing interaction leading to superconductivity, astonished the scientific community [18–20].

The large variety of properties highlights the key role of the $LaAlO_3/SrTiO_3$ interface as a model compound to understand the new physics at complex oxide interfaces. The coexistence of phases confined to a two-dimensional layer highlights the importance of a precise understanding of the $LaAlO_3/SrTiO_3$ interface on the route towards future applications. However, only little is known about the exact mechanisms that lead to the formation of the observed interfacial electronic properties. Widely-accepted models accurately describe the emergence of the conducting two-dimensional electron gas, but do not give a complete picture of the underlying phenomena and microscopic physics. In particular, this includes the localization of charge carriers [21,22]; growth-induced defects, such as oxygen vacancies [23]; and the stabilization of local inhomogeneities [17] at the interface. Hence, the question whether an extrinsic or intrinsic mechanism yields the conducting state at the $LaAlO_3/SrTiO_3$ is still a highly debated topic in the literature [24–29]. To clarify this key question, detailed knowledge about the electronic structure at the interface of the pathway to the formation of the two-dimensional electron gas is required.

The optical Second Harmonic Generation (SHG) process is essentially the generation of a light wave with the frequency 2ω by doubling the frequency ω of the incident wave [30–32]. The measured SHG signal, i.e., the intensity of the double-frequency wave, is proportional to the absolute square of the second order susceptibility tensor $\chi^{(2)}$. The SHG levels diagram and the basic idea of the experimental layout is given in Figure 1 together with the labeling of the axes and polarization directions. The model which describes the process will be addressed in the following section. Here, we would like to focus first on the most relevant point: because of the very thin region which is affected by the 2DEG, one needs a powerful and non-invasive technique specifically sensitive to thin interfaces, and SHG is ideal with respect to these issues. The reason for this lies in the sensitivity of SHG to the symmetry of the system. A very simple and general argument shows that, under the dipole approximation, the SHG signal vanishes in any centro-symmetric medium, i.e., when the system is insensitive upon inverting each ion position from the point $\vec{r}$ to the point $-\vec{r}$. This strong symmetry can be fully achieved, evidently, only if the material extends to an infinite size in space. The presence of a surface geometrically hampers the inversion symmetry, regardless of any property of the material, and, therefore, any interface (a surface is nothing but an interface with vacuum) acts as a source of SHG signal [33]. One can regard the issue also by looking at the alteration of the arrangement of the ions in the bulk; the breaking of the inversion symmetry in a condensed matter system corresponds directly to a shift of ionic and electronic charges, and the SHG probes the resulting polarization. It is very important to understand that, for the above mentioned reasons, there is no pre-determined "probing depth" of SHG, as there is for many other surface techniques, such as ARPES and others. The optical wavelengths typically employed in experiments are in the 100 nm to the micrometer scale, and the penetration length could be, depending on the specific material, up to meters or more. This is irrelevant as the bulk of the centro-symmetric material will be illuminated by the incident light, but it will not emit any signal. The emitting region of the SHG signal instead covers the region affected by the symmetry breaking, i.e., the region where the material differs significantly by its bulk structure.

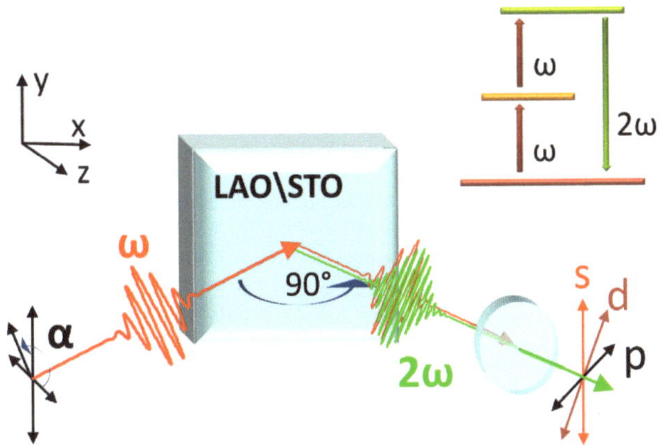

Figure 1. Experimental scheme and polarization geometry. The input fundamental light with frequency ω is converted to its Second Harmonic by 90° reflection on the sample surface. The generic input polarization angle is α while the output polarization angles are fixed to be parallel to the incident plane (p), to the sample surface (s), or at 45° (d). In the upper panel, the SHG energy levels diagram is shown.

The main objective of this review is an experimental in-depth analysis of the optical non-linearities at the $LaAlO_3/SrTiO_3$ interface as driven by the electronic structure and its modification with increasing number of $LaAlO_3$ mono-layers. This is achieved by exploiting the light-polarization, spectral, spatial, and temporal degrees of freedom of SHG. Throughout this review, fundamental questions such as the role of localized carriers, the presence and influence of structural inhomogeneities, and the carrier dynamics across the interface are addressed. In particular, this includes the notorious discussion about the extrinsic or intrinsic origin of the two-dimensional electron gas at the $LaAlO_3/SrTiO_3$ interface. Hence, the experimental results here presented enhance the understanding of the puzzling mechanisms yielding two-dimensional conduction, as well as the temporal dynamics involved, whose time scales are of fundamental importance for possible future applications, such as optical devices with ultrashort switching times. In the following, we will use the abbreviations LAO and STO for $LaAlO_3$ and $SrTiO_3$, respectively.

2. Theoretical Background

We report here, for the sake of clarity the first part of the SHG theory developed in Ref. [34], and we refer the reader to that work for further details.

Under very general circumstances, the dielectric polarization density, i.e., the dipole moment per unit volume $\overrightarrow{P}(t)$ can be expanded as the electrical field power series of the incoming wave $\overrightarrow{E}(t)$. A simplified expression for this, that terminates the Taylor expansion to the third term, is the following:

$$\overrightarrow{P}(t) \sim \varepsilon_0 \left(\hat{\chi}^{(1)} \overrightarrow{E}(t) + \hat{\chi}^{(2)} \overrightarrow{E}^2(t) + \hat{\chi}^{(3)} \overrightarrow{E}^3(t) \right) \tag{1}$$

where $\varepsilon_0 \simeq 8.85 \times 10^{-12}$ F·m^{-1} is the dielectric constant in vacuum and $\hat{\chi}^{(n)}$ are the expansion coefficients, also known as non-linear susceptibilities. The non-linear polarization $\overrightarrow{P}(t)$ will act as a source of electromagnetic radiation in the wave equation. We are interested particularly in the second term, where the second power of a field oscillating at frequency ω will clearly give an oscillation of the non-linear polarization at double frequency 2ω, and, therefore, a far-field wave E_{SHG} oscillating at this frequency. Equation (1) does not show the tensorial nature of the non-linear susceptibilities $\hat{\chi}^{(n)}$ explicitly. Assuming the oscillatory behavior of second polarization term $\overrightarrow{P}^{(2)}(t)$ and the electric field $\overrightarrow{E}(t)$ with complex amplitudes $P(\omega)$ and $E(\omega)$, respectively, let us rewrite the second term alone, assuming the sum over repeated indices:

$$P_i^{(2)}(2\omega) = \varepsilon_0 \chi_{ijk}(2\omega) E_j(\omega) E_k(\omega) \tag{2}$$

where the apex (2) has been removed on the tensor components for the sake of simplicity. It is seen in Equation (2) that if we apply the inversion operator on centro-symmetric crystals, all vectors (as they are true vectors and not pseudo-vectors) are changing their sign, while the material-related susceptibility does not (because of the inversion symmetry), so that we obtain $P^{(2)}(2\omega) = -P^{(2)}(2\omega)$ which can hold true only if the SHG signal is vanishing, as already discussed in the Introduction.

To be useful, the SHG technique must establish a clear interpretive link between the measurable $\hat{\chi}^{(2)}$ "macroscopic" tensor (or its variations) and the microscopic quantities of interest, for example in terms of electronic band occupation, orbital reconstructions, etc. Regardless, before entering this complex problem, we should first recognize that, in case of LAO/STO as well as any other interface, there are two possible sources of signal, i.e., two distinct interfaces, the LAO/STO and the LAO/air (or vacuum), and the signals from them can also interfere as the sub-wavelength distance of the two regions ensures a strong coherence in any given experimental condition. In the case of LAO/STO hetero-structures, the SHG signal, comes predominantly from the LAO/STO interface and it is generated mainly on the STO side of the interface [35]. It should be not a surprise, therefore, that a non-vanishing SHG signal was found also in the bare STO substrate, and its characteristics are essential—as we will discuss in the following sections—as a reference for evaluating the alterations induced by the presence of a LAO overlayer, instead of air.

The intensity of the SHG signal is the actually measured quantity in SHG experiments. Depending on the particular experiment, we can change the parameter which drives the changes in the SHG signal, we record SHG Spectroscopy as a function of the incident photon energy $\hbar\omega$, SHG Imaging as a function of the spatial coordinates on the sample surface, SHG Polarimetry as a function of the input/output waves polarizations, and so forth, not to mention external parameters, such as temperature or bias electromagnetic fields. In every case the signal is proportional to the square of the reflected SHG electric field, $I_{\text{SHG}} \propto |E_{\text{SHG}}|^2$. Equation (2) refers to the electromagnetic fields inside the material, but we can only access the fields outside of it. Therefore, we need to re-write that equation keeping into account the Fresnel tensors $\hat{L}^{in,out}$ which couple the electric fields of the incident and emitted light to those which are present inside the interface. By also taking into account the time dependence of the fields and after a straightforward calculation, we have:

$$E_{\text{SHG}} = \frac{i\omega E_0^2}{\varepsilon_0 c \cos\beta} \chi_{\text{eff}} \tag{3}$$

where $c \simeq 2.99 \times 10^8$ m/s is the speed of light in vacuum, E_0 is the amplitude of the impinging wave, β is the angle of incidence, and χ_{eff} is defined as follows:

$$\chi_{\text{eff}} = \int e_i^{\text{out}} L_{ii}^{\text{out}} \chi_{ijk} L_{jj}^{\text{in}} L_{kk}^{\text{in}} e_j^{\text{in}} e_k^{\text{in}} dz, \tag{4}$$

where $e_{i,j,k}$ are the optical polarization unit vectors in a vacuum. This approach is valid as long as the integration thickness is much smaller than the wavelength.

We now introduce a Cartesian reference system having the interface plane oriented along the xy plane and the z-axis perpendicular to it, in such a way that the incident plane is xz. We assume that both input and output waves are linearly polarized, with an arbitrary polarization angle α with respect to the incidence plane. The electric field unit vectors are then the following:

$$\begin{aligned} \mathbf{e}^{\text{in}} &= (\cos\alpha\cos\beta, \sin\alpha, \cos\alpha\sin\beta) \\ \mathbf{e}^{\text{out}} &= (-\cos\alpha\cos\beta, \sin\alpha, \cos\alpha\sin\beta) \end{aligned} \tag{5}$$

At this point it is very convenient to introduce a short notation for specific polarizations: p for a direction parallel to the plane of incidence ($\alpha = 0$), s parallel to the interface plane ($\alpha = \pi/2$), and d oriented about 45° between p and s ($\alpha = \pi/4$).

As mentioned, we can neglect the contribution of the LAO–air interface, and so the three non-zero components of the Fresnel tensor **L** can be approximated by the following expressions

$$\begin{aligned} L_{xx} &= \frac{2\cos\beta'}{n\cos\beta + \cos\beta'} \\ L_{yy} &= \frac{2\cos\beta}{\cos\beta + n\cos\beta'} \\ L_{zz} &= \frac{2\cos\beta}{n(n\cos\beta + \cos\beta')} \end{aligned} \tag{6}$$

where n is the refractive index of the SHG-active interface, the STO substrate in this case and β' the propagation (refraction) angle inside the medium. Let us note that the refractive index is dependent on the light frequency. Therefore, it must be calculated at ω (2ω) when considering the input (output) wave.

At room temperature, $SrTiO_3$ crystallizes in the ABO_3 cubic perovskite structure. The presence of a (001)-cut surface/interface lowers the symmetry to have a four-fold rotation around the surface normal and two vertical mirror planes, so that the space group it belongs to is the $4mm$ in the Hermann–Mauguin notation, or C_{4v} in the Schoenflies notation, which possesses five different irreducible representations, named A_1, A_2, B_1, B_2, and E. When considering the material electronic states associated with these symmetry representations, the first four identify singlet states and the last a doublet. A_1 is the fully symmetric state, which behaves like 1, z or z^2; A_2 is odd with respect to both σ_v and σ_d inversion planes and behaves like the rotation matrix R_z; B_1 is odd with respect to $\pi/2$ rotations and σ_d reversals, and behaves like $x^2 - y^2$; B_2 is odd with respect to $\pi/2$ rotations and σ_v reversals and behaves like xy, and the doublet E behaves like the variables x, y or the products xz, yz [36].

The allowed tensor component of the non-linear susceptibility for the $4mm$ group are:

$$\begin{gathered} \chi_{zzz} \\ \chi_{zxx} = \chi_{zyy} \\ \chi_{xxz} = \chi_{yyz} = \chi_{xzx} = \chi_{yzy} \end{gathered} \tag{7}$$

By substituting Equations (5)–(7) in Equation (4) we obtain the expression of χ_{eff} for the noticeable polarization combinations shown in Figure 1 p-input p-output (pp), s-input p-output (sp), and d-input s-output (ds),

$$\begin{aligned}
\chi^{\text{eff}}_{sp} &= \chi_{zxx}\delta L^{\text{out}}_{zz}(L^{\text{in}}_{yy})^2 \sin\beta \\
\chi^{\text{eff}}_{ds} &= \chi_{xxz}\delta L^{\text{out}}_{yy} L^{\text{in}}_{yy} L^{\text{in}}_{zz} \sin\beta \\
\chi^{\text{eff}}_{pp} &= \chi_{zzz}\delta L^{\text{out}}_{zz}(L^{\text{in}}_{zz})^2 \sin^3\beta + (\chi_{zxx} d L^{\text{out}}_{zz} L^{\text{in}}_{xx}) - 2\chi_{xxz} d L^{\text{out}}_{xx} L^{\text{in}}_{zz}) L^{\text{in}}_{xx} \sin\beta\cos^2\beta,
\end{aligned} \tag{8}$$

where δ denotes the effective thickness of the polar interface and the χ_{ijh} are space-averaged across this thickness. Now the choice of the input/output polarization can be understood; only one element of $\hat{\chi}^{(2)}$ is present in sp and ds expression, and namely χ_{zxx} and χ_{xxz} respectively, while the third element χ_{zzz} cannot be experimentally isolated, but it can be calculated starting from experimental data after proper subtraction of the sp and ds components. The sets of sp, ds, and pp polarizations provide the full set of tensor elements; although it is not the only choice, it is the most simple and straightforward for the experiments. It is worth noting that a similar calculation has been carried out for other crystal orientations as well and, namely, (110), symmetry group $mm2$ in Hermann–Mauguin notation or C_{2v} in the Schoenflies notation and (111), symmetry group $3m$ or C_{3v}. The explicit calculations are present in Ref. [37], together with the SHG experimental results on those particular crystal orientations, we will briefly discuss these non-standard orientations in a dedicated section of this review and we remind the reader to that reference for details.

When discussing the resonance structure of $\hat{\chi}^{(2)}$, one has first to make very general assumptions, including that (i) the bulk system is perfectly centro-symmetric, (ii) the surface/interface induced symmetry breaking only occurs along the z direction, and (iii) the effect of the symmetry breaking is not too large, i.e., it can be evaluated within the dipole approximation. These points imply that the parity is broken along z but not along x and y, so that the interfacial electronic states can be either even or odd by inversion of the x,y coordinates. Moreover, we recognize that the full $4mm$ symmetry holds only in the center of the Brillouin zone (Γ point), whereas in other points it is partially broken by the crystal momentum $k \neq 0$. Finally, we remark that any resonant SHG can be of three types: resonant at the fundamental photon energy, resonant at the doubled photon energy, and resonant at both energies. Having all this in mind, one additional ingredient is needed before discussing the SHG resonances, the material band-structure.

3. LAO/STO Structure and Properties

The bulk optical gap of STO is known to be associated primarily with a transition from oxygen $2p$ to titanium $3d$-T_{2g} orbitals. Its typical perovskite crystal structure is sketched in Figure 2. The Ti^{4+} ions are sixfold coordinated by O^{2-} ions, while each of the Sr^{2+} ions is surrounded by four TiO_6 octahedra, highlighted in green in the figure. Thus, each Sr^{2+} is coordinated by 12 O^{2-} ions. Inside the TiO_6 octahedra, while a hybridization of the O_{2p} states with the Ti_{3d} states leads to a pronounced covalent bond, Sr^{2+} and O^{2-} have an ionic bonding character. Hence, $SrTiO_3$ has mixed ionic-covalent bonding properties and this leads to a unique structure, which makes it a model electronic material. For a given planar direction, for example (100), which is the most commonly used direction for growing LAO/STO hetero-structures, there are always two distinct types of equally spaced alternating atomic planes, with different arrangements of the two components, namely SrO-planes and TiO_2-planes, sketched in Figure 2b,c. A transition from cubic to tetrahedral symmetry occurs if the temperature is lowered below approximately 100 K, or if a foreign cation/dopant is introduced into the lattice. The distortions are attributed to three main effects, dimensional effects, deviations from the ideal composition and the Jahn-Teller effect. An important example, which will be discussed in detail in the next sections, belongs to the second class. Oxygen-deficient strontium titanate is formed if the valence of the Ti cation is changed, either by heat treatment in an oxidizing/reducing atmosphere or by doping in the Sr sub-lattice. The oxygen content varies accordingly and the oxygen vacancies are sorted preferentially with respect to the local structure, i.e., a tetrahedral coordination. Non-stoichiometric thin layers of LAO can alter the ionic organization too via proximity effects, which lead to non-trivial behavior of the 2DEG at the interface. The second order

Jahn–Teller effect is responsible for other types of distortions of the perovskite structure, namely a ferroelectric distortion of the cations at the Sr site, but this issue will not be addressed here.

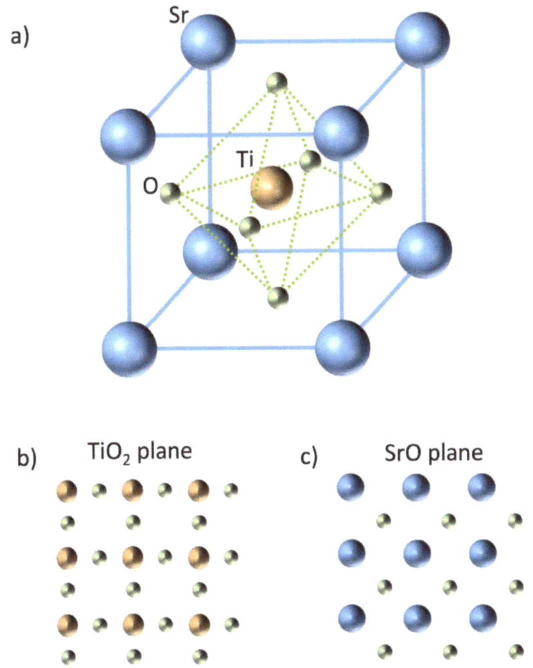

Figure 2. (**a**) 3D $SrTiO_3$ perovskite unit cell. The solid lines represent the cubic cell borders while the dashed lines represent the TiO_6 octahedral cage which highlights the Ti-O coordination. (**b**,**c**) 2D view of the titanium oxide (**b**) and strontium oxide (**c**) planes.

In its stoichiometric form, $SrTiO_3$ is an insulator with an indirect band gap of 3.2 eV (at $T = 0$ K) separating the valence band and conduction band. The six-fold coordination of Ti ions from the surrounding oxygen cage creates a crystal field which splits the degenerate states of Ti$-3d$ by 2.4 eV between T_{2g} and e_g symmetries. Figure 3a shows a sketch of the electronic structure where the cation has electron configuration d_0. The valence band which corresponds to the highest occupied molecular orbitals (HOMO) is composed mainly of oxygen orbitals 2s and 2p and the conduction band which corresponds to the lowest unoccupied molecular orbitals (LUMO) is mainly cationic, arising from empty Ti-d states. The gap between the HOMO and LUMO states makes $SrTiO_3$ a band insulator. The Sr cations in $SrTiO_3$ are, in general, strongly electro-positive and, therefore, play a minor role in the electronic structure. In any case, they play an important role in modifying the TiO_6 connectivity of the perovskite structure and, thus, the electronic structure, but this happens mainly because of their physical size, i.e., the fraction of the unit cell volume they occupy. Indeed, the presence of "intrinsic defects", such as ionic vacancies or structural displacements can lead to modifications of the electronic structure and electronic conductivity of the material, just in a similar way to extrinsic defects, such as dopants. Adding vacancies or doping elements in the $SrTiO_3$ lattice creates defects with an effective charge relative to the host lattice. In this review, we will focus mainly on two different source of intrinsic and extrinsic effects, and namely oxygen vacancies and Nb^{5+} (in replacement of Ti^{4+} ions) dopants. The former plays a crucial role in the discussion about the origin of the 2DEG, as for many years the exact role of oxygen vacancies, which are to some extent very difficult to control, generated a long-lasting scientific dispute. The latter has been chosen as a typical source of extrinsic doping, to be compared with the intrinsic vacancy-induced conduction and with the interface-induced 2DEG.

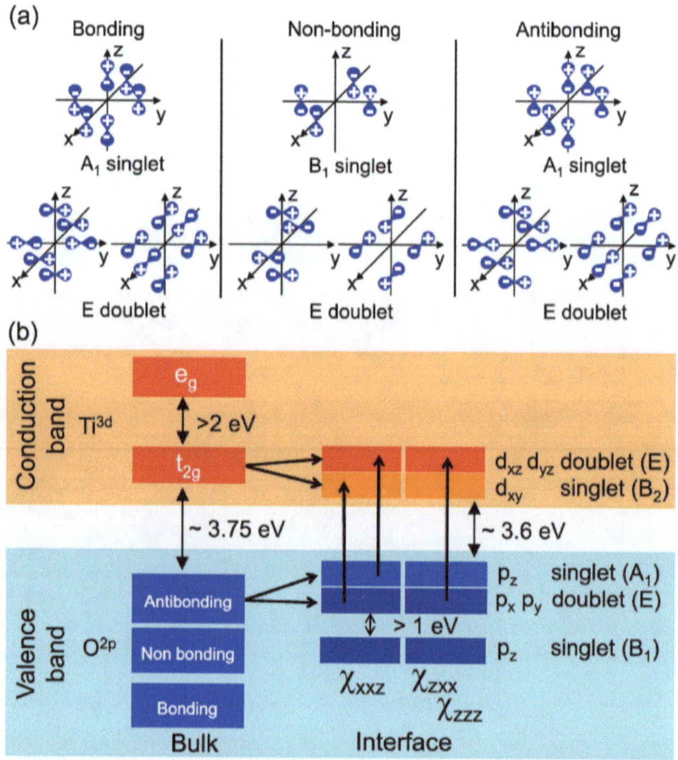

Figure 3. (**a**) Different arrangements of O (2*p*) orbitals forming the highest occupied molecular orbitals (HOMO) are indicated. In the bulk, they are grouped in triply degenerate levels according to the cubic $m3m$ symmetry, but at the surface the symmetry is broken (local $4mm$ symmetry) and the degeneracy between the singlet and doublet states is removed. (**b**) The O(2*p*)–Ti(3*d*) 2ω transitions allowed for all three symmetry components χ_{ijk} within the symmetry group $4mm$ according to Equation (7). Reprinted with permission from Ref.[38].

4. Experimental Methods

The experimental methods used to prepare the samples and to perform the SHG measurements are different for every work covered in this review. Nonetheless, in this section we will give a very brief overview on the most common and widely used techniques and we would like to remind the interested readers to the cited references for more details. Thin $LaAlO_3$ films are grown on $SrTiO_3$ usually by Pulsed Lare Deposition (PLD) technique, assisted by high-energy electron diffraction (RHEED) oscillations, used to monitor the epitaxial growth and to count the number of deposited monolayers. Typical growth temperature and base oxygen pressure can significantly vary, but usual values are about $\sim$800 °C and 10^{-4} mbar. In fewer cases, the growth technique of choice is Molecular Beam Epitaxy (MBE) with substrate temperatures in the range 700–800 °C. Typical values of interfacial sheet conductance are $\sigma_S = 10^{-5}$– 10^{-4} Ω^{-1} at 300 K for samples above conduction threshold. The typical light-source for the experiments here reported is a Ti:Sa laser with ultrashort pulses ranging between 35 and 130 fs, with pulse energies in the range of the few μJ and fluences around 1–10 mJ/cm^2. For spectroscopy experiments, the laser fundamental light at 800 nm is converted by an Optical Parametric Amplifier to any wavelength in the range 250–2500 nm. The optical setup depends on the specific experiment, of course, but, in general, the emerging SHG signal is collected in reflection geometry, for several reasons. First of all, the presence of a thick bulk substrate may contribute to the signal via electric quadrupole, hampering the effort of achieving an interface-only sensitive signal. On the other hand, the back side is usually not optically polished and it is covered by silver glue, which is necessary during the sample growth. Therefore, the glue should be removed after growth, with the risk of contaminating the interface side of the sample. The fundamental light is finally filtered out by optical filters, and the SHG is usually detected by means of photo-tubes or, for imaging experiments, projected with photographic objectives onto a high-sensitivity liquid-nitrogen-cooled digital CCD camera.

5. SHG Spectroscopy

Figure 4 shows the *pp*, *ds*, and *sp* spectra for different LAO coverage n ranging from 0 to 12 mono-layers in the energy range 3.2–4.2 eV. Between 1.5 and 3.2 eV all spectra are flat and featureless and, therefore, this energy range is not very informative from a spectroscopic point of view and it is not shown in the figures. On the other hand, energies above 4.2 eV are difficult to access experimentally. A marked increase in SHG yield of about 3.6 eV is visible in all samples, and this energy corresponds to the edge of the direct O(2p)→Ti(3d) band-gap transition in STO [35,39]. However, the main peak is found at around 3.8–4.0 eV. Looking to the ellipsometry data found in the cited literature [40,41], this energy corresponds to a maximum of the dielectric function. Such a clear spectroscopic feature in the SHG spectra shows that the SHG probes the STO electronic structure and it is driven by the O(2p) and Ti(3d) orbitals involved in the local surroundings of the interface. Another important piece of information is obtained by the comparison of different SHG spectra as a function of n, where n is the number of LAO over-layers. The integrated SHG yield is small for $n = 0$ and $n = 1$ and it becomes slightly larger for $n = 2$. Regardless, at $n = 3$, a sudden and substantial boost in all components and across the spectral range takes place. Let us note that different $n = 3$ samples can exhibit a really strong sample-to-sample variability. Let us remark here that the onset of 2D conduction at the interface happens only when $n \geq 4$ and none of the $n = 3$ samples is found to be conductive, including those with high SHG yield. The observation of a threshold value for a discontinuous structural transition preceding the onset of conduction was first reported in 2009 by Savoia and co-authors [42] in the seminal paper "Polar catastrophe and electronic reconstructions at the interface $LaAlO_3/SrTiO_3$: Evidence from optical second harmonic generation" and has since been confirmed by other works [43]. In order to visualize the sample-to-sample fluctuations at $n = 3$, only two samples showing maximum and minimum SHG yield are labeled, respectively, "3+" for the highest SHG signal (similar to samples with $n > 3$) and "3−" for the lowest signal (comparable to samples with $n < 3$). The origin of such fluctuations at $n = 3$ is still unknown at present, but it is not surprising that, in the proximity of a phase transition, the reconstruction threshold is particularly sensitive to slight changes in the growth conditions. For a larger amount of LAO layers, the SHG signal is stable with only a minor increase with respect to the $n = 3+$ sample, indicating that the interface is fully reconstructed. Again, no major changes of SHG are found between $n = 3+$ and $n = 4$, i.e., the threshold for bidimensional conduction. SHG spectra have been taken at low temperature, in the range from 10 K to 300 K, as well as with different capping materials ($NdGaO_3$—NGO—and $LaGaO_3$—LGO—in place of LAO) [38]. In the low-temperature regime, SHG is enhanced in all components and all samples. In particular, the band-edge 3.8 eV peak in *pp* and *ds* spectra experiences a larger enhancement than other spectral components for both STO and LAO/STO samples. More importantly, at very low temperatures (∼10 K), the number of resonances revealed by the SHG spectra exceeds the number of those identified by the SHG selection rules. This is due to the energy shifts of different sub-bands, as reported by angle resolved photoemission spectroscopy (ARPES) at similar temperatures [44]: the degeneracy of the d_{xz}, d_{yz}, p_x, and p_y bands, is lifted and ARPES data indicate that the SHG peak splitting of about 120 meV is in good agreement with this picture. The mechanisms are still not clear, but possible candidates are the electron-confinement effect, spin–orbit coupling, and/or low-temperature tetragonal/orthorhombic distortions. Regarding the different capping layers, different topmost materials share the same qualitative behavior. This confirms that SHG is probing the electronic inter-band transitions of STO rather than any specific transition of the polar overlayer, whose effect is only visible through the changes it induces in the STO underneath. However, a few quantitative differences with respect to the specific overlayer materials can be cited, including (i) the SHG signal of the LAO/STO samples is always larger than that of the other two interfaces, (ii) in the *ds* and *pp* spectra of conductive samples the high-energy part is less enhanced in LGO/STO and NGO/STO samples than in LAO/STO, and (iii) in the *pp* spectra of insulating samples the overall SHG yield of LAO/STO is larger than that of the other samples.

SHG Spectroscopy has been the most important tool in order to address the problem of the microscopic origin of SHG in these materials and its relationship with the physical effect under investigation, i.e., the 2D insulator-to-metal phase transition. SHG indirectly senses the interfacial charge asymmetry. This asymmetry may be induced by a structural reorganization of the interface (in particular by Ti ions displacement) or by the presence of an electric field. In the second case, for instance, when a charged region is created close to the interface, the consequent interfacial electric field polarizes all the orbitals involved in the SHG emission and when the charge injection occurs, the SHG signal is enhanced. Therefore, the fact that the SHG signal jumps at $n = 3$ means that charges start to be injected one layer "before" the onset of conduction, and thus they must be localized. They contribute to the interfacial field build-up, but not to transport. If the injected charges accumulated exactly at the interface, they would provide a perfect screening of the LAO polarization so that there would be no evidence of electric-field build-up in STO. On the contrary, the injected charges diffuse over a few unit cells because of kinetic energy, and create a charge gradient and, consequently, an electrostatic field that spatially decays in STO [45]. Regardless, it may seem, at a first glance, surprising that the emergence of conduction is completely separated from the orbital reconstruction. Possible solutions for this conundrum are (i) SHG is less sensitive to mobile charge than localized ones; (ii) the $n = 4$ LAO layer promotes to conduction some of the localized carriers at $n = 3$, so that the total carrier density does not change; and (iii) the conductive carrier density is much smaller than the localized one.

The most important result of the SHG spectroscopy is summarized in Figure 3, where the assignment of each orbital transition to a specific element of the tensor $\chi^{(2)}$ is shown. The observed transitions are very robust as they are the source of SHG for all studied parameters. The $4mm$ symmetry was found to be generally preserved, albeit with minor but interesting deviations. Furthermore, the overall enhancement of the low-temperature SHG signal, and consequently an increase in polarity at the interface, supports the picture of a spatial displacement of the titanium ions, which are prone to move at lower temperatures due to the quantum para-electric nature of STO. Slight deviations from this general picture, as dictated by the symmetry selection rules, are (i) the suppression of the 3.6 eV peak in the *sp* spectra is not always complete, showing that the symmetry is partially lifted due to a distortion of TiO_6 octahedra; and (ii) the SHG signal is related to the in-plane lattice mismatch of the LAO/STO, NGO/STO and LGO/STO insulator interfaces, demonstrating that substrate distortions are probably induced or at least modulated by strain in the epitaxial film. These results clearly prove that fine details in the interfacial structure are reflected in, and revealed by, SHG spectroscopy. However, where this detailed sensitivity is most striking is in the study of the influence of substrate termination on LAO/STO electronic properties. The 2D electron gas is actually observed only for the TiO_2-terminated STO substrate (n-type doping, TiO_2/LaO interfaces), while for the SrO-terminated substrate (SrO/AlO_2 interfaces) a p-type conductive interface is predicted based on the polar catastrophe scenario, but the samples are found to be insulating for any LAO coverage and growth condition [6,7].

SHG Spectroscopy in this case shows a clear qualitative difference between the spectra of TiO_2 and SrO terminated samples, as can be observed in Figure 5. The 3.8–4.0 eV peak is strongly suppressed in the *sp* spectrum of SrO terminated interfaces with respect to TiO_2 ones. In Ref. [39], a detailed resonance analysis has been performed and the the model presented in Figure 6 summarizes the resulting findings, as explained in the following.

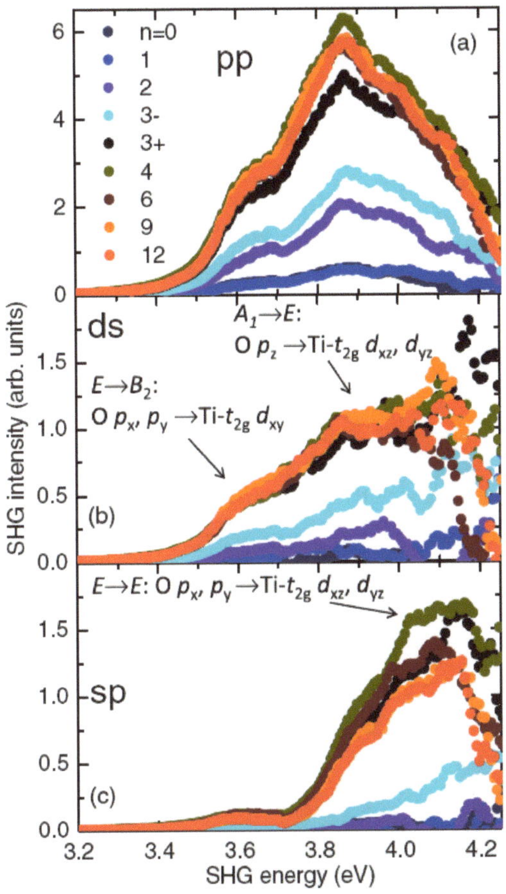

Figure 4. SHG spectra of LAO/STO hetero-structures for (**a**) *pp*, (**b**) *ds*, and (**c**) *sp* polarization configurations with different LAO thickness *n*. Samples with $n = 3$ show strong sample-to-sample variation in SHG yield. This value of *n* represents a threshold for a discontinuous interfacial reconstruction which anticipates the conduction, i.e., the presence of free charge carriers. Here, the two extreme cases have been labeled with "+" or "−" in order to distinguish them, so that the "3+" sample is the one with the highest SHG signal, and "3−" the one with lowest SHG signal. These spectra cover the direct O(2*p*)→Ti(3*d*) band-gap transition of STO, which appears as an increase in SHG intensity at approximately 3.6 eV. The symmetry-based SHG selection rules take into account that the lowest energy transition at 3.6 eV is present in the spectrum of *pp* and *ds*, but not in *sp*. The same arbitrary scale is used in all three panels. Reprinted with permission from Ref. [38].

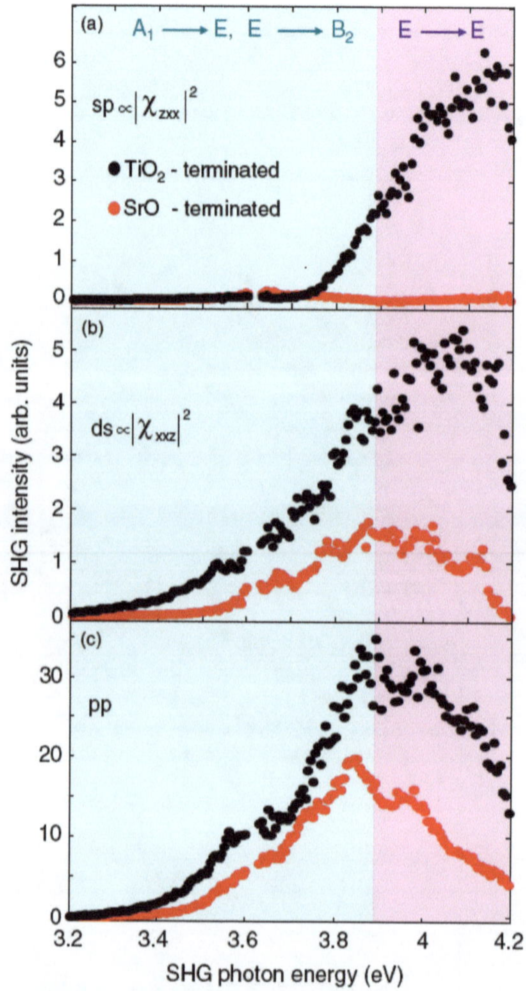

Figure 5. Comparison of SHG spectra from LAO/STO hetero-structures with five LAO unit cells and different interfacial atomic terminations. The labels *sp* (**a**), *ds* (**b**), and *pp* (**c**) refer to different combinations of $\omega/2\omega$ polarizations (see text). The spectra for the SrO-terminated substrate (red symbols) are multiplied by a factor of 4 for better visibility. The colored areas divide the graph into two energy ranges corresponding to the identified electronic transitions: the blue area is dominated by $A_1 \rightarrow E$, $E \rightarrow B_2$ transitions, while the purple area is driven by E→E transitions. Reprinted with permission from Ref. [39].

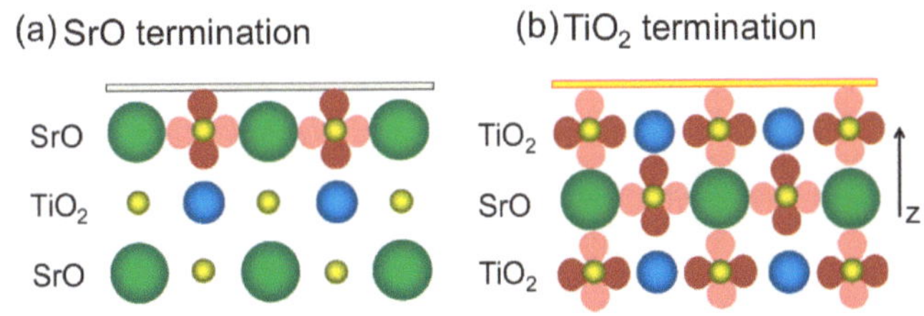

Figure 6. A schematic image of the two possible STO terminations. The $2p$ oxygen orbitals are shown in dark red and pink, where dark red highlights direction maximizing overlap with the Ti atom. In the case of SrO-terminated STO $2p_z$ orbitals point towards the underlying Ti atoms, thus favoring orbital overlap. Reprinted with permission from Ref. [39].

The almost complete suppression of the 3.8–4.0 eV peak in the sp spectrum is attributed to a suppressed transition $E \rightarrow E$: $\text{O}(p_x, p_y) \rightarrow \text{Ti-}T_{2g}(d_{xz}, d_{yz})$. Since this cannot be explained by the selection rules, as both terminations share the same interfacial symmetry, it must be assumed that in the SrO interface the SHG is mainly generated in the orbitals located within a single atomic plane from the interface, as these are strongly polarized due to structural proximity effects. Only the oxygen p_z orbitals have significant overlap with the Ti($3d$) orbitals of the underlying TiO_2 layer, so that only transitions involving p_z should be pronounced in the spectrum. Thus, the transition $E \rightarrow E$: $\text{O}(p_x, p_y) \rightarrow \text{Ti-}T_{2g}(d_{xz}, d_{yz})$ of the sp spectrum is suppressed in SrO-terminated systems. Furthermore, the transitions $A_1 \rightarrow E$: $\text{O}(p_z) \rightarrow \text{Ti-}T_{2g}(d_{xz}, d_{yz})$ remain possible, and contribute to the non vanishing ds and pp spectra. However, here the $E \rightarrow B_2$: $\text{O}(p_x, p_y) \rightarrow \text{Ti-}T_{2g}(d_{xy})$ transition will be suppressed. Conversely, in TiO_2 terminated samples, the charge injection that drives the electronic reconstruction at the interface creates a space charge region spanning a few unit cells, as already mentioned above, thus displacing the Ti ions and developing an electric field that polarizes all the orbitals involved in the SHG emission. Thus, the overall SHG yield will be higher and all symmetry-allowed $O(2p) \rightarrow Ti(3d)$ transitions will contribute to the SHG spectrum. The suppression of the sp signal in the SrO-terminated samples is strong evidence that only the highest STO atomic plane at the interface is significantly polarized, which, in turn, implies that no significant charge injection occurs in the SrO-terminated samples, unlike the TiO_2-terminated samples. These findings have helped ruling out some of the proposed models for explaining the lacking of p-type doping in LAO/STO hetero-structures, and, in particular, all those based on charge carrier trapping.

6. Non-Standard LAO/STO Hetero-Structures

As opposed to the case of different terminations where, according to the polar catastrophe model, the SrO-terminated interface is expected to be conductive and is in fact insulating, there is the case of amorphous-LAO structures in which no polar build-up can be postulated since the polar LAO layer is grown on the STO substrate in an amorphous phase and yet electric conduction at the interface is observed. Additionally, in this case, SHG was found to be capable of helping to address the apparently inconsistent observation [46]. Figure 7 shows a temperature scan of the SHG signal from amorphous-LAO (a-LAO, panel a) and crystalline-LAO (c-LAO, panel b) upon heating (red datapoints) and cooling (blue datapoints). It is also worth noting that during the measurement, the temperature was held at 110 °C for half an hour, for reasons explained below. It was already shown by transport measurements that, contrary to c-LAO structures where conductivity is recovered almost 100% after any number of thermal cycles, in the case of a-LAO the 2D electron gas is lost after the very first cycle, indicating an extrinsic origin of the doping charge, and, in particular, pointing to oxygen vacancies, for the simple reason that heating the sample in air has the same effect as filling the oxygen vacancies at the interface, which, in turn, can be formed by ion bombarding during PLD growth of the samples. The SHG results in Figure 7 confirm that the doping mechanisms that induce the interfacial polarity detected by SHG are very different in the two material systems. For the c-LAO/STO sample, the behavior of the SHG signal as a function of temperature is qualitatively the same as that observed for the conductivity. The SHG signal is approximately constant during both the heating and cooling phases. After the cooling step, the SHG signal from the c-LAO/STO interface recovers approximately its initial value at room temperature. In the case of a-LAO/STO, a pronounced decrease in the SHG signal is observed, analogous to the conductivity. However, SHG also exhibits some significant features which are not visible otherwise. The decrease in the SHG signal already begins at about 110 °C: above this temperature, the SHG signal decreases spontaneously at an approximate rate of about 30% per hour. This spontaneous change does not take place for conductivity, which starts to decrease only above 200 °C. After this break, the temperature is increased again by 10 °C every five minutes. In addition, the relative strength of different SHG components are irreversibly changed by heating; the *ds* signal decreases more rapidly than the *sp* one, until it vanishes at about 300 °C, while the other two components never become zero. Below about 250 °C, *ds* is always greater than *sp*, while this relationship is reversed when above 250 °C. Finally, the SHG signal remains constant at the values reached for the highest temperatures upon cooling. Therefore, the *ds* signal remains lower than that of *sp*, thus reversing the initial signal hierarchy of these two polarization combinations. This result shows that the obtained insulating interface is different from the untreated single crystal STO surface since the latter generates an *sp* signal which is always lower than that of *ds*, provided that the energy of the SHG photon is not close to the optical resonances of STO. The naive picture of an oxygen-free interface which after heating and subsequent oxygenation becomes identical to the pristine strontium titanate surface is probably true, but at least partially misleading or incomplete. The most remarkable observation is that the different evolution of SHG and electrical conductance under a thermal treatment clearly indicates two different donor mechanisms at the a- and c-LAO/STO interfaces, oxygen vacancies in the former case and electron reconstruction driven by interfacial polar discontinuity in the second case. The question then arises naturally, why is a similar SHG signal observed for all the c- and a-LAO/STO hetero-structures?

As already mentioned, SHG indirectly probes the charges and their spatial distribution at the interface. When a space-charge region is created at the interface, this develops an electric field, $E_{polar}(z)$, that polarizes the electronic orbitals involved in the SHG process. This contribution may be accounted by the third term of Equation (1), where one field is given by $E_{polar}(z)$ and the other two are the optical field at ω. Note that the structural contribution to SHG is still described by the second term of Equation (1). By assuming $\overline{\chi}^{(3)}$ slowly varying over the thin layer d_{polar} affected by the electronic reconstruction one finds that

$$\overline{\chi}_{\text{eff}} \propto \overline{\chi}^{(2)} + \overline{\chi}^{(3)} \int E_{polar}(z)dz = \overline{\chi}^{(2)} + \overline{\chi}^{(3)} V_{well}, \tag{9}$$

where the bar indicates the spatial averaging over the z coordinate. The variation of $\overline{\chi}_{\text{eff}}$ at fixed $\overline{\chi}^{(2)}$ is therefore proportional to the depth V_{well} of the potential well induced by the charge distribution at the interface. An estimate of V_{well} as a function of LAO thickness is provided in Figure 8 for both the a-LAO/STO and c-LAO/STO interfaces, with an average of the polarization combinations. All crystalline samples above the conductance threshold have an interfacial V_{well} similar to amorphous samples in the saturation region. This result indicates the existence of a universal depth of the interfacial potential well, despite the fundamentally different doping mechanism acting in these two material systems. A possible microscopic scenario for the case of the a-LAO/STO interface, which could help explain this similarity, is based on the hypothesis that a $\delta-$doping mechanism caused by oxygen vacancies takes place at the a-LAO/STO interface. The band diagram of a-LAO/STO would, therefore, be determined by the combination of two basic elements, i.e., the typical "dip" of the bands formed in the vicinity of a δ-doping sheet and the well-known misalignment of the conduction band of LAO and STO, making the states on the LAO side of the δ-doped interface inaccessible to electronic wave functions. The interaction of these two components would lead to a band diagram such as the one in Figure 8c, showing significant similarities, on the STO side, to the one predicted by the electron reconstruction model for the *c*-LAO/STO case, as shown in Figure 8b. One possible explanation is that this depth is actually dominated by the difference between the minimum energy of the conduction band in the STO bulk and the Fermi level, a difference which should be largely independent of the properties of the interface.

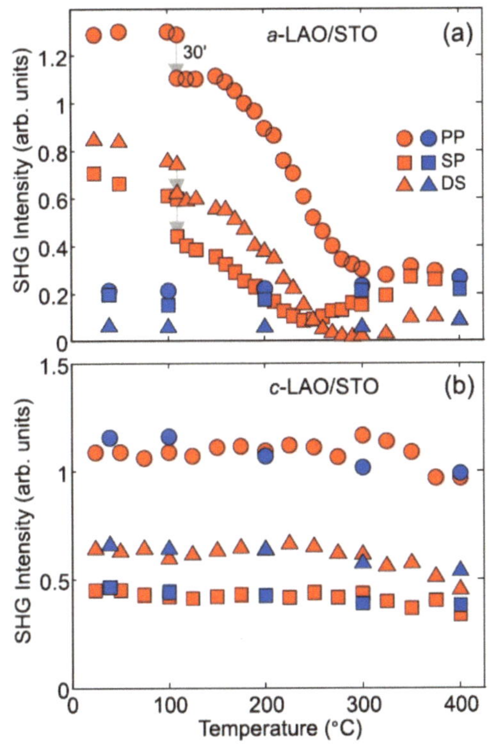

Figure 7. SHG intensity as a function of temperature for 10 equivalent-unit-cell thick *a*- and *c*-LAO/STO samples (panel (**a**,**b**), respectively), for three polarization combinations: *pp* (circles), *ds* (triangles), and *sp* (squares). Red (blue) points refer to the heating (cooling) stage. The arrow at 110 °C indicates the beginning of the SHG signal decrease. At this temperature a pause of about 30 min (30′ label in the figure) is performed before continuing the temperature scan. Reprinted with permission from Ref. [46].

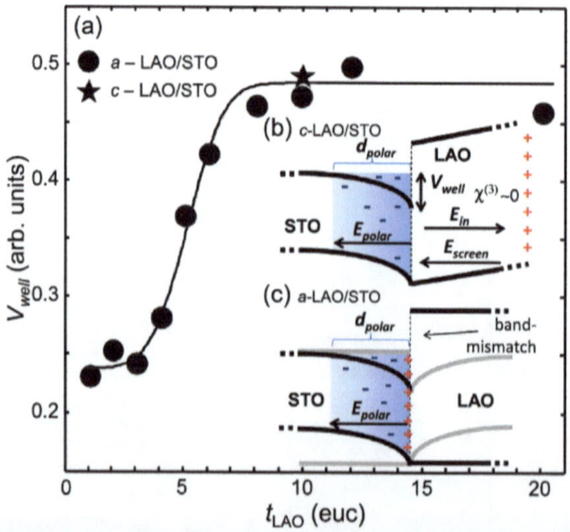

Figure 8. In panel (**a**) the values of the potential-well depth V_{well} for *a*-LAO/STO (circles) and *c*-LAO/STO (star) extracted from the SHG signal are reported as a function of LAO thickness t_{LAO} (in euc, i.e. equivalent unit-cell). The solid line is just a guide for the eyes. In the insets (**b**,**c**) the possible band diagrams of an *c*-LAO/STO and *a*-LAO/STO interface are shown. For the *c*-LAO/STO interface, the bending of the LAO bands is due to the built-in electric field, E_{in}, as predicted by the polar catastrophe model. At the critical thickness, this field is offset by E_{screen}. The band structure of the *a*-LAO/STO interface is tentatively represented as a typical potential well derived from a δ-doping. The latter implies that the electronic states on the LAO side (represented by gray wings) are rendered inaccessible by the mismatch of the Fermi level between STO and LAO. The latter is highlighted by the gray lines on the STO side, which show the band structure in the absence of δ-doping. Reprinted with permission from Ref. [46].

Another interesting case of non-standard LAO/STO hetero-structure is the one with unbalanced chemical composition. In particular, Ref. [47] shows that samples with nominal formula $La_{(1-\delta)}Al_{(1+\delta)}O_3$ were grown by molecular beam epitaxy (MBE), and only Al-rich samples were found to be conductive. In Ref. [48], SHG has been deployed to study such interfaces, and the main result is shown in Figure 9, where different batches of samples grown at different temperatures and with different chemical composition are compared. The signal of the samples grown at about 720 °C is only slightly increasing in the Al-rich region of the diagram, while the other two groups show a much more pronounced increase. The sample richer in Al, in particular, shows an increase in SHG of about 50% compared to that of the 755 °C set with the same Al/La ratio, while the conductivity for these samples is approximately the same or even slightly higher for the set grown at 755 °C. The most remarkable thing is that the conductivity of the 720 °C samples saturates by increasing the Al content, but SHG does not demonstrate the same behavior. In contrast, the 755 °C sample set shows similar saturation behavior to the corresponding conductivity. In both cases, the doping level at which SHG starts to increase for the 755 °C and 790 °C set correspond to the start of conduction (indicated by the red and green arrows in Figure 9). The reason for the lack of conduction in the La-rich samples was proposed by first-principles calculations in Ref. [47]; in the La-rich films the excess was not found to substitute on the Al sites of the LAO crystal because it is too large, forming Al_2O_3-vacancy complexes instead. Conversely, in Al-rich films the Al is incorporated in place of the La sites. This leads to few cation vacancies in Al-rich films vs. many cation vacancies in La-rich films, which results in a significantly higher diffusion coefficient for cations in the La-rich films. In turn, this makes it possible to offset the diverging electric potential via cations migration. On the contrary, in the Al-rich films, the low diffusion coefficient of cations forces an electronic reconstruction in order to compensate the diverging electric potential, thus leading to the formation of a 2D electron gas. Conversely to conductivity, at the higher substrate temperature, the SHG signal does not saturate, since an increase in Al atoms is directly accompanied by the formation of new active SHG defect dipoles. The same behavior is not observed at lower substrate temperatures, probably due to the ability of excess Al to hinder crystallization in the $La_{(1-\delta)}Al_{(1+\delta)}O_3$ film. Under Al-rich growth conditions when the growth temperature is low, a large volume of the $La_{(1-\delta)}Al_{(1+\delta)}O_3$ film is amorphous and Al excess in this amorphous matrix does not produce defect dipoles. On the other hand, increasing the substrate temperature counteracts this glass-forming tendency. A substrate temperature of about 790 °C is sufficient to maintain the $La_{(1-\delta)}Al_{(1+\delta)}O_3$ crystalline film everywhere so that Al substituting onto the La sites may produce defect dipoles. This explains the anomalous behavior of SHG as a function of Al/La ratios for the samples grown at the highest temperature.

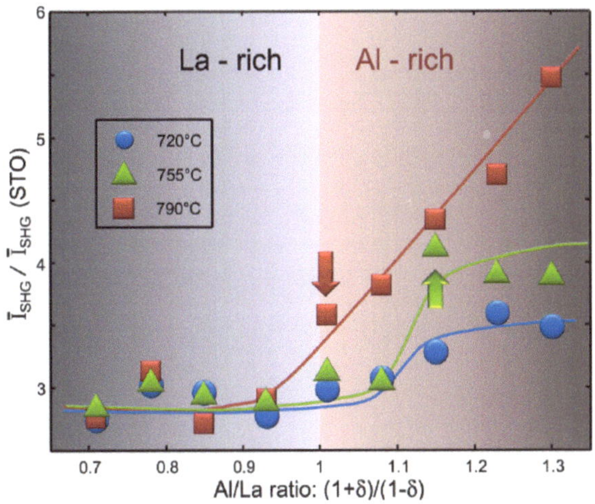

Figure 9. The total yield of SHG as a function of the Al/La ratio, normalized to the value found for the STO–air interface. The two arrows indicate the onset of conduction for the 790 °C-set (red) and the 755 °C-set (green) samples. The shaded areas denote the two non-stoichiometric chemical compositions, respectively, rich in La and Al. The shade gradient shows the direction of increase in the chemical component in each respective area. Solid lines are a guide for the eyes. Reprinted with permission from Ref. [48].

Another example of the SHG technique applied to non-standard LAO/STO heterostructures is provided by interfaces with different in-plane orientations, the LAO/STO (110) interface is an interface formed by the deposition of LAO atomic layers on oriented STO substrates (110). According to the polar catastrophe scenario, the LAO/STO interface (110) is formed by a sequence of $[O_2]^{4-}[LaAlO]^{4+}$ planes over $[O_2]^{4-}[SrTiO]^{4+}$ and thus should not be conductive. Conversely, the LAO/STO interface (111) also displays a polar discontinuity, being formed by an alternative sequence of $[Al]^{3+}[LaO_3]^{3-}$ over $[Ti]^{4+}[SrO_3]^{4-}$ planes, and thus should be practically comparable to standard LAO/STO(001) interfaces. In fact, all these interfaces show the same insulator-to-metal transition. At first sight, this result may appear to challenge the polar catastrophe picture. However, recent theoretical studies [49] show that at the LAO/STO (110) interface the STO surface is not an ideal stoichiometric surface, but the ground state is characterized by a buckled TiO termination. This predicted ionic structural distortion leads again to an interfacial polar discontinuity, that, in turn, drives the electronic reconstruction through the same mechanism of the (001) interface. In order to confirm or disprove this particular theoretical prediction, SHG was employed in Ref. [37] as the ideal experimental tool to probe the existence of a ionic interfacial polarity at the (110) interface. Figure 10 shows an example of the SHG Polarimetry performed on LAO/STO (111) samples. A detailed symmetry study on (001), (111), and (110) orientations has demonstrated that the theoretical picture of Ref. [49] is indeed correct, and, in particular, that the interfacial polarity is already very strong before the onset of conduction for the (110) samples, with no appreciable signs of multi-polar contributions, and this comes directly to the surface structural distortion of the STO (110) substrate, which undergoes the formation of a buckled TiO termination.

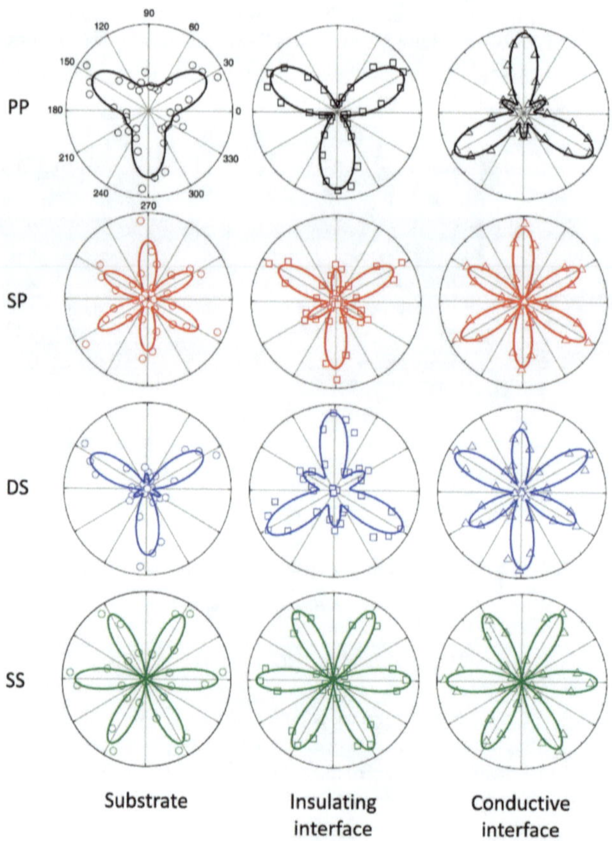

Figure 10. SHG Polarimetry: the signal is shown as a function of the azimuthal angle ϕ and polarization combinations (*pp*, *sp*, *ds*, *ss*) for LAO/STO(111) samples. Open symbols are data and the solid lines are best-fit curves according to the 3 m symmetry group. Reprinted with permission from Ref. [37].

7. SHG Imaging

The possible presence of local inhomogeneities in the electronic structure of the interface was proposed [50–53] to account for the delayed onset of conductivity at $n = 4$ instead of $n = 3$, as measured with non-transport techniques, such as SHG. In addition to the problem of free/bound charges, magnetic [13], and chemical [54] phase separation has been reported too. SHG was found to be one of the very few techniques able to address these points, because of its ability to investigate the in-plane spatial distribution of the interface reorganization, through the lateral degrees of freedom in the focal plane. SHG can be, in simple words, used to make a "polar picture" of the sample. The resulting picture will not represent the morphological properties of the surface, but rather the strength of the polar nature of the buried interface. In Ref. [55], the SHG Imaging technique has been used in order to address this issue, and quite surprisingly it was found that all heterostructures with $n \neq 1$, including the substrate itself ($n = 0$), are very homogeneous, while only for a single mono-layer of LAO coverage, distinct inhomogeneities are visible, which are characterized by the formation of bright regions with a lateral extension of $\simeq$30 μm. The case of $n = 1$ seems to be quite special. Electronic and chemical inhomogeneities have been found to agree with the observations, although the latter is less plausible, while oxygen vacancies clustering was rejected for its failure to account for the absence of any observable structure at $n \geq 2$. In Ref. [56], SHG Imaging has been proposed as a tool for non-invasive, non-destructive, real-time, in situ imaging of oxide epitaxial film growth. Figure 11 shows the proposed idea and the preliminary measurements which are showing the feasibility of the device. In the upper panel, a simplified sketch of a Pulsed Laser Deposition (PLD) setup is shown equipped with the standard Reflection High Energy Electron Diffraction (RHEED) setup and the new proposed SHG imaging tool. The films can be monitored during growth with lateral resolution of $\leq$1 μm on a very large area (about 1 cm^2). The potential of the device is demonstrated by an ex situ analysis of thin epitaxial $SrTiO_3$ films grown on (110) $NdGaO_3$ substrates, taken as prototypical example of a perovskite-based hetero-interface, such as LAO/STO and many others. Of course the real future device should be able to work on a larger variety of materials in order to be fruitfully implemented in real PLD chambers, but it can be taken as a proof of principle. The example picture shown in panel Figure 11a indicates the presence of large regions with pronounced polar nature, which have been independently observed by Piezo-response Force Microscopy (PFM) in order to investigate the electrostatic potential of the sample surface. In addition, a 180° phase shift in SHG occurs when the order parameter of a broken-symmetry is reversed, as happens for instance in the case of adjacent ferroelectric (FE) domains with opposite polarization. The possibility to investigate the SHG signal phase is a relatively easy task which can be obtained by interference between the signal from the target sample and that of a reference sample, and, thus, this makes it possible to distinguish between FE domains having the same size of FE polarization, but reversed orientation. The SHG imaging provides complementary information to the well-established in situ RHEED and it can reveal otherwise elusive in-plane inhomogeneities of electrostatic, chemical, or structural nature. We are not going here to review the entire field of in situ SHG monitoring because it goes beyond the scope of the present review work, but we remind the reader to the comprehensive review on this specific topic in Ref. [57] and references therein.

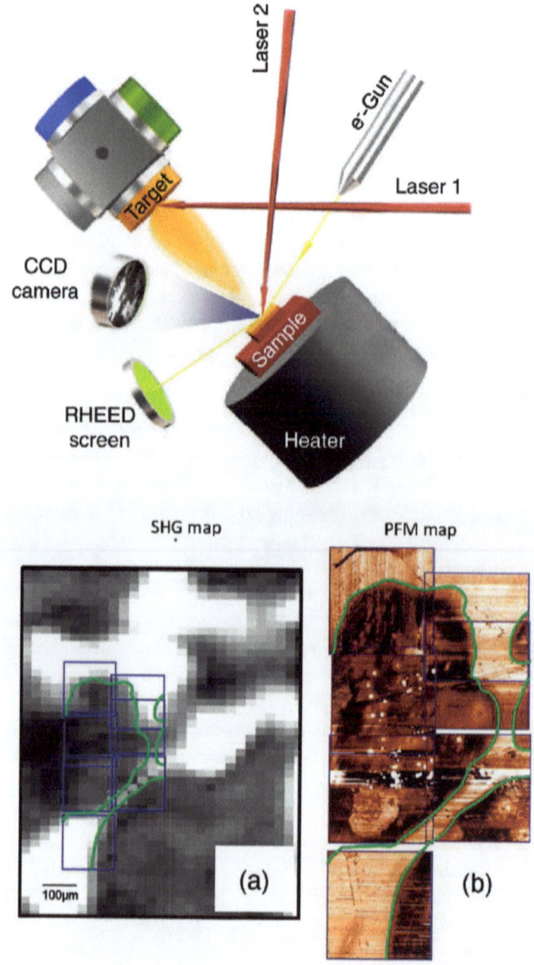

Figure 11. **(Upper Panel)**: Schematic layout showing a possible device concept for in situ monitoring of epitaxial film growth by SHG. The usual geometry of the PLD chamber is not affected by the presence of an additional laser beam since both the source and the detector could be located outside the growth chamber. **(Lower panels)**: (**a**) SHG intensity map on a $SrTiO_3/NdGaO_3(110)$ sample and (**b**) PFM image of the same area showing the electrostatic potential measured in contact mode (bright regions correspond to a higher electron density). The colored lines are guides for the eye. Reprinted with permission from Ref. [56].

8. Time-Resolved SHG

The last topic we would like to cover in this review is the dynamic behavior of charge carriers in LAO/STO. So far, only the static state at the interface has been discussed, although fast and ultra-fast dynamics would introduce a powerful degree of freedom for understanding the nature of charge carriers. In Ref. [58], a pump-probe scheme is used in which a pump beam with photon energy of 4.35 eV, i.e., above the direct band gap of STO, can excite photo-carriers which, in turn, induce a change in interfacial polarity. The latter is probed by the SHG of a second infrared beam as completed in standard static experiments. The results of these pump-probe experiments show that the polarization state can be optically enhanced or attenuated in picoseconds. A physical model based on the effects of charge propagation at the interface and transient polarization build-up is proposed to explain these observations. Time-dependence allows, in fact, to separate the various interactions that contribute to the interfacial ground state across the different time scales on which they respond to optical excitation. For example, transient absorption spectroscopy shed light on the strong influence of LAO film on self-trapped polarons at the LAO/STO interface [59]. The SHG dynamic study revealed the mechanisms driving the accumulation and depletion of light-induced polarity at the interface: the authors found an induced sub-picosecond optical change of interfacial polarity greater than 50%. This surprising effect is driven by the competition between three different mechanisms, screening, asymmetric drift, and trapping. Taking Equation (9) into consideration again, the dynamic quantity measured in the pump-probe mode can be considered as follows:

$$\overline{\chi}^{(2)}(t) = \overline{\chi}^{(2)} + \Delta\overline{\chi}^{(2)}(t) \tag{10}$$

where t indicates the time-delay between the pump and the probe pulses, the bar indicates the spatial averaging over the z coordinate. The time-dependence is fully included in the second term of the sum, whose components, according to Equation (9), can be written as:

$$\Delta\overline{\chi}_{ijk}(t) = \Delta V(t)\overline{\chi}^{(3)}_{ijkl} \tag{11}$$

In other words, it is assumed that the main dynamical effect is captured by the time evolution of the quantum well energy depth $V(t) = V_0 + \Delta V(t)$ induced by the interfacial electric field. Here, the authors refer to $\overline{\chi}_{loc,ext}$ in order to indicate the χ_{xxz} and χ_{zxx}, respectively. This distinction highlights the more "localized" or "extended" nature of the electronic transitions identified for those two components according to the more pronounced nature of the respective final states (d_{xy} for localized and $d_{xz,yz}$ for extended) [51]. Figure 12a and the zoomed-in view in Figure 12b are showing the dynamical behavior of the "extended" component χ_{zxx} for two representative samples with $n = 2$ (LS2) and $n = 6$ (LS6) LAO coverage. The SHG signal shows a pronounced and ultrafast drop within less than 1 ps, followed by a quick recovery (on the 100 ps time scale) and a subsequent, and much slower restoration, of the ground state (in several nanoseconds). What is striking and interesting is the fact that both the substrate and the insulating interface (STO and LS2 in the figure) present an "overshooting", meaning that the polarity becomes larger than that of the ground state after the initial ultrafast drop, so that the recovery to unexcited level of the signal happens in the positive range (polarity decreases to unexcited level), while for the case of conductive sample (LS6) the initial drop is much more pronounced and the recovery happens completely in the negative range (polarity increases to unexcited level). Exploiting a comprehensive investigation about the change in linear reflectivity, LAO coverage, and photo-excitation energy density, a relatively simple model, depicted in the sketches of Figure 13, has been proposed to interpret the complex dynamics of these electron systems. It is seen that the dynamics is the result of several different and competing mechanisms, whose microscopic origins will be roughly explained in the following, with the help of the four sketches of Figure 13. Panel (a) shows the band bending at the interface in the ground state and highlights the more localized nature of the d_{xy} orbitals (purple) compared to the more extended nature of the $d_{xz,yz}$ orbitals (green); panel (b) shows the drift of the photo-carriers in the presence of a pre-existing equilibrium electric field ("screening drift" mechanism), whose effect is mainly captured by the ultrafast drop of the SHG signal within the first picosecond; panel (c) shows the so-called photo-Dember effect, i.e., a transient and local electric field originated by the anisotropic diffusion of electron vs. holes which induces a net shift of the electron cloud with respect to the hole cloud ("charge-propagation" mechanism); finally panel (d) shows the interfacial charge carrier trapping (and subsequent recombination) as the process contributing to the increased interfacial polarity ("transient polarization" mechanism). In order for the latter mechanism to obtain the observed evidence, the enhanced interfacial polarity, it is worth noting that the charge sign of the trapping centers must be positive for inducing an electric field concordant to the existing quantum-well field. It is necessary to point out that this mechanism would work equally well in the case of other kind of charge accumulation, such as, for instance, the formation of large polarons at the interface, although charge trapping is the most likely scenario, in particular because of the well-known presence of oxygen vacancies (positive trap sites). By the study of the fit function parameters derived from this model, and by comparison with literature [60], it was possible to demonstrate that polarons are good candidates to explain the observed transition state. The lattice deformation accompanying the polaron state must break the inversion symmetry to be visible in SHG, as occurs near the interface where the photo-generated polarons are trapped by defects and become polar. The most important evidence to this conclusion can be drawn by the comparison with the model parameters and the known drift mobility of polarons in STO, and this demonstrates how much important the dynamical studies are, even if the ground state, rather than the excited states, is the main target of the investigation.

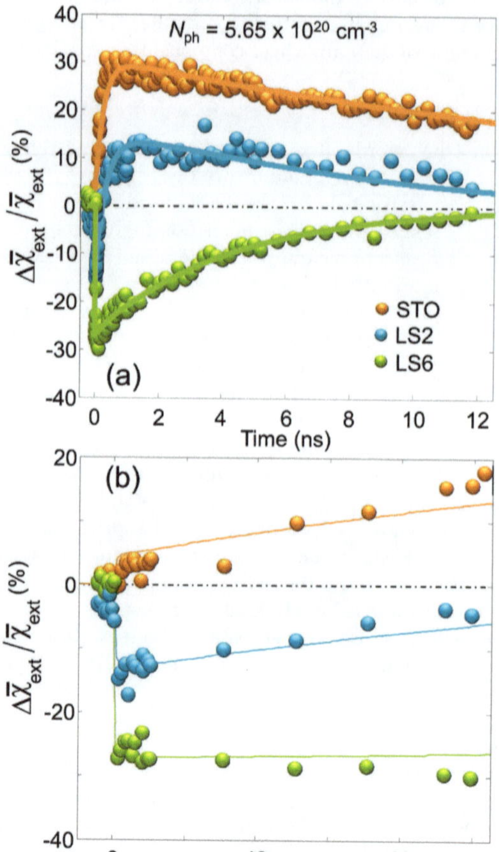

Figure 12. (**a**,**b**) Temporal evolution of $\Delta\overline{\chi}_{ext}/\overline{\chi}_{ext}$ for a pump photon density $N_{ph} = 5.65 \times 10^{20}$ cm^{-3}. The label *ext* indicates the χ_{zxx} component, while LS stays for LAO/STO and the number indicates the LAO overlayer unit cells. Solid lines are best-fit curves according to the dynamical model proposed. Note the striking difference of the dynamics between insulating (orange and blue points) and conductive (green points) samples. Reprinted with permission from Ref. [58].

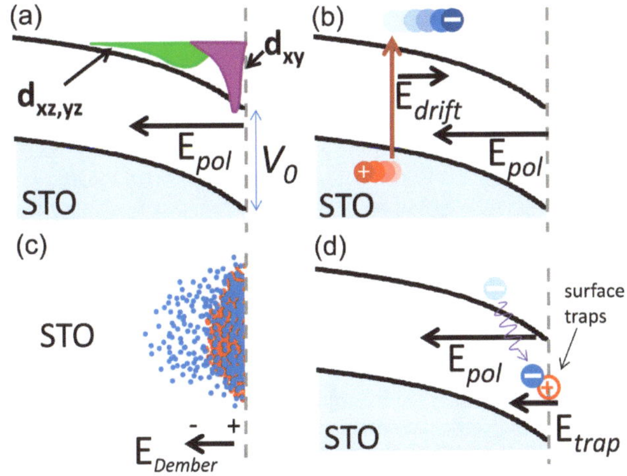

Figure 13. Non-equilibrium charge dynamics near the LAO/STO and air/STO interfaces. (**a**) Band bending and different space depths of charge concentration associated, respectively, with the d_{xy}-like and $d_{xz,yz}$-like sub-bands of STO at the interface. We note that the medium beyond the dotted line can be air, as in bare STO, or a LAO overlay. (**b**) Screening drift as the first mechanism inducing the interfacial polarity change. (**c**) Photo-Dember effect as a second mechanism inducing the interfacial polarity variation. (**d**) Interfacial charge trapping as a mechanism inducing a transient polarization. Reprinted with permission from Ref. [58].

9. Conclusions

In this work, we have reviewed more than 10 years of experiments applying SHG to the investigation of oxide hetero-structures with a particular focus on the textbook LAO/STO interface. The results clearly show that SHG is capable of providing complementary information on these material systems compared to the standard tools that are widely used in material science. The advantages of SHG compared to the latter are many-fold: (i) SHG can reach vertical spatial resolution on the subnanometric scale; (ii) it is a contact-free technique; (iii) it can directly sense interfacial polarity; (iv) it can be used in situ and real-time for monitoring the hetero-structure growth without disturbing the particle trajectories; and (v) it can reach temporal resolution on the femtosecond time-scale.

Let us summarize here what we believe are the main results in the field. SHG spectroscopy revealed a reorganization of the electronic structure at the interface at $n = 3$ which precedes the formation of the 2DEG at $n = 4$. This reorganization is invisible to other techniques which are sensitive to the conductive charges only. Its extreme symmetry breaking sensitivity made it possible to observe directly the spectral influence of the atomic termination of the $SrTiO_3$ substrate and the influence of extrinsic carrier doping, especially at low temperature. Exploiting the lateral spatial degrees of freedom of SHG, the transverse plane homogeneity of the interface is revealed. Lateral inhomogeneities on a length scale of $\approx$30 μm are found in samples with $n = 1$ epitaxial mono-layer of $LaAlO_3$ (and absent in $n > 1$ samples), due to a phase-separation involving domains with different electronic and chemical properties. Studies on different chemical composition (stoichiometry of the components) and crystalline structure (crystalline vs. amorphous hetero-structures) helped to single out the role of so-called intrinsic effects (polar catastrophe) from the extrinsic effects (oxygen vacancies and others). SHG time-resolved experiments on the femtosecond-to-nanosecond time-scales revealed a stunningly complex carrier dynamics after optical excitation, which can be regarded as the result of a competition between at least three different dynamical relaxation channels. These results shed light on the occupancy and lifetime of electronic states, and suggest that light pulses can be exploited to enhance or attenuate the polarization at the interface and possibly the charge density and/or mobility.

On the other hand, it can be difficult to directly link the SHG signal to the microscopic properties of the interface under study, so that often the interpretation of the SHG results remains confined to a qualitative level. Enhancing the diagnostic and predictive capacities of SHG would require a major effort on the microscopic theoretical modeling of SHG, thus allowing it to compete with other powerful tools in material science, as photo-emission spectroscopy or scanning tunnel microscopy. Notwithstanding these limitations, some of the results here summarized have proved that SHG is able to provide valuable information on the physics of these interfaces that cannot be easily obtained with other techniques. The main strength of SHG is its capacity of sensing all the charges injected at the interface, both localized or mobile. In the case of the STO-based interfaces it is clear that SHG is more sensitive to the first kind of charges, thus allowing to reveal a phenomenon that acts as a precursor for the onset of conductivity. This precursory interfacial reconstruction, due to charges that are injected but localized at the interface, cannot be of course highlighted by the electrical measurements routinely used to characterize these interfaces. This is even more true if we consider LAO/STO interfaces with different terminations of the (100) STO substrate. In this case, there are few techniques that may provide a significant comparison between these two interfaces, and SHG is one of the simplest. The results obtained by SHG spectroscopy in this particular case are simply striking. First, they show that SHG is able to provide useful information on the level of a single atomic layer. Second, they prove that, at the SrO terminated LAO/STO interface, charges are not injected, an important piece of information, not possible to obtain with other techniques.

Moreover, SHG can directly measure the interfacial polarity and even the interfacial electric field that builds-up on charge injection. In this respect, SHG spectroscopy has unequivocally demonstrated that $SrTiO_3$ is a fundamental ingredient for the formation of the 2DEG gas at these interfaces. Injected charges spread in a thin layer on the STO side and there they build-up a polar electric field leading to the formation of a confining quantum-well. SHG can measure directly the depth of this quantum-well, thus allowing, for instance, to single out universal characteristics of fundamentally different systems as amorphous and crystalline LAO/STO interfaces and to call into question once again the fundamental role played by $SrTiO_3$ in the physics of these material systems.

In addition, SHG spectroscopy as a function of temperature has shown that the polar asymmetry induced by the build-up of the polar electric-field is always accompanied by a structural reorganization of the interface due to the displacement of Ti ions that are very much prone to shift from their equilibrium position because of the paraelectric nature of $SrTiO_3$. This is another important piece of the puzzle representing the physics of these material systems, since it has been lengthily debated if the charge injection induces a structural or electronic reconstruction of the interface; SHG results show that both are in place.

In conclusion, this review article has provided several evidences to convince the readers that Second Harmonic Generation may be an invaluable experimental tool for investigating buried interfaces (LAO/STO and other oxide hetero-structures) which share common chemi-physical characteristics. In this conclusive section we have summarized the major results that, in our opinion, have been obtained by means of SHG and which make this technique unique in comparison with other standard tools used in this field. Along with the review, we have delivered several examples highlighting the richness of information that SHG can gather, with the aim of providing a good starting point for further developments in this fruitful research field.

Funding: This research received no external funding.

Conflicts of Interest: The authors declare no conflict of interest.

References

1. Orton, J.W. *The Story of Semiconductors;* OUP Oxford: Oxford, UK, 2004.
2. Sze, S.; Ng, K.K. *Physics of Semiconductor Devices*, 3rd ed.; Wiley: Hoboken, NJ, USA, 2007.
3. Lee, H.; Christen, H.; Chisholm, M.; Rouleau, C.; Lowndes, D. Strong polarization enhancement in asymmetric three-component ferroelectric superlattices. *Nature* **2005**, *433*, 395. [CrossRef]
4. Norton, D.; Chakoumakos, B.; Budai, J.; Lowndes, D.; Sales, B.; Thompson, J.; Christen, D. Superconductivity in $SrCuO_2$-$BaCuO_2$ superlattices - Formation of artificially layered superconducting materials. *Science* **1994**, *265*, 2074. [CrossRef]
5. Takahashi, K.; Kawasaki, M.; Tokura, Y. Interface ferromagnetism in oxide superlattices of $CaMnO_3/CaRuO_3$. *Appl. Phys. Lett.* **2001**, *79*, 1324. [CrossRef]
6. Ohtomo, A.; Hwang, H. A high-mobility electron gas at the $LaAlO_3/SrTiO_3$ heterointerface. *Nature* **2004**, *427*, 423. [CrossRef]
7. Nakagawa, N.; Hwang, H.; Muller, D. Why some interfaces cannot be sharp. *Nat. Mater.* **2006**, *5*, 204. [CrossRef]
8. Thiel, S.; Hammerl, G.; Schmehl, A.; Schneider, C.; Mannhart, J. Tunable Quasi-Two-Dimensional Electron Gases in Oxide Heterostructures. *Science* **2006**, *313*, 1942. [CrossRef] [PubMed]
9. Huijben, M.; Rijnders, G.; Blank, D.; Bals, S.; Aert, S.V.; Verbeeck, J.; Tendeloo, G.V.; Brinkman, A.; Hilgenkamp, H. Electronically coupled complementary interfaces between perovskite band insulators. *Nat. Mater.* **2006**, *5*, 556. [CrossRef] [PubMed]
10. Caviglia, A.D.; Gariglio, S.; Reyren, N.; Jaccard, D.; Schneider, T.; Gabay, M.; Thiel, S.; Hammerl, G.; Mannhart, J.; Triscone, J.M. Electric field control of the $LaAlO_3/SrTiO_3$ interface ground state. *Nature* **2008**, *456*, 624. [CrossRef] [PubMed]
11. Cen, C.; Thiel, S.; Hammerl, G.; Schneider, C.; Andersen, K.; Hellberg, C.; Mannhart, J.; Levy, J. Nanoscale control of an interfacial metal-insulator transition at room temperature. *Nat. Mater.* **2008**, *7*, 298. [CrossRef]
12. Rijnders, G.; Blank, D. An atomic force pencil and eraser. *Nat. Mater.* **2008**, *7*, 270. [CrossRef]
13. Bert, J.; Kalisky, B.; Bell, C.; Kim, M.; Hikita, Y.; Hwang, H.; Moler, K. Direct imaging of the coexistence of ferromagnetism and superconductivity at the $LaAlO_3/SrTiO_3$ interface. *Nat. Phys.* **2011**, *7*, 767. [CrossRef]
14. Cancellieri, C.; Fontaine, D.; Gariglio, S.; Reyren, N.; Caviglia, A.; Fête, A.; Leake, S.; Pauli, S.; Willmott, P.; Stengel, M.; et al. Electrostriction at the $LaAlO_3/SrTiO_3$ interface. *Phys. Rev. Lett.* **2011**, *107*, 056102. [CrossRef] [PubMed]
15. Reyren, N.; Thiel, S.; Caviglia, A.; Kourkoutis, L.F.; Hammerl, G.; Richter, C.; Schneider, C.; Kopp, T.; Rüetschi, A.S.; Jaccard, D.; et al. Superconducting Interfaces Between Insulating Oxides. *Science* **2007**, *317*, 1196. [CrossRef] [PubMed]
16. Brinkman, A.; Huijben, M.; van Zalk, M.; Huijben, J.; Zeitler, U.; Maan, J.; van der Wiel, W.; Rijnders, G.; Blank, D.; Hilgenkamp, H. Magnetic effects at the interface between non-magnetic oxides. *Nat. Mater.* **2007**, *6*, 493. [CrossRef]
17. Xie, Y.; Bell, C.; Yajima, T.; Hikita, Y.; Hwang, H. Charge Writing at the $LaAlO_3/SrTiO_3$ Surface. *Nano Lett.* **2010**, *10*, 2588. [CrossRef]
18. Dikin, D.; Mehta, M.; Bark, C.; Folkman, C.; Eom, C.; Chandrasekhar, V. Coexistence of Superconductivity and Ferromagnetism in Two Dimensions. *Phys. Rev. Lett.* **2011**, *107*, 056802. [CrossRef]
19. Li, L.; Richter, C.; Mannhart, J.; Ashoori, R. Coexistence of magnetic order and two-dimensional superconductivity at $LaAlO_3/SrTiO_3$ interfaces. *Nat. Phys.* **2011**, *7*, 762. [CrossRef]
20. Kalisky, B.; Bert, J.; Klopfer, B.; Bell, C.; Sato, H.; Hosoda, M.; Hikita, Y.; Hwang, H.; Moler, K. Critical thickness for ferromagnetism in $LaAlO_3/SrTiO_3$ heterostructures. *Nat. Commun.* **2012**, *3*, 922. [CrossRef]
21. Lee, J.; Demkov, A. Charge origin and localization at the n-type $SrTiO_3/LaAlO_3$ interface. *Phys. Rev. B* **2008**, *78*, 193104. [CrossRef]
22. Zhou, K.J.; Radovic, M.; Schlappa, J.; Strocov, V.; Frison, R.; Mesot, J.; Patthey, L.; Schmitt, T. Localized and delocalized Ti 3*d* carriers in $LaAlO_3/SrTiO_3$ superlattices revealed by resonant inelastic x-ray scattering. *Phys. Rev. B* **2011**, *83*, 201402. [CrossRef]
23. Chen, Y.; Pryds, N.; Kleibeuker, J.; Koster, G.; Sun, J.; Stamate, E.; Shen, B.; Rijnders, G.; Linderoth, S. Metallic and Insulating Interfaces of Amorphous $SrTiO_3$-Based Oxide Heterostructures. *Nano Lett.* **2011**, *11*, 3774. [CrossRef] [PubMed]
24. Siemons, W.; Koster, G.; Yamamoto, H.; Harrison, W.; Lucovsky, G.; Geballe, T.; Blank, D.; Beasley, M. Origin of charge density at $LaAlO_3$ on $SrTiO_3$ heterointerfaces: Possibility of intrinsic doping. *Phys. Rev. Lett.* **2007**, *98*, 196802. [CrossRef]
25. Herranz, G.; Basletić, M.; Bibes, M.; Carrétéro, C.; Tafra, E.; Jacquet, E.; Bouzehouane, K.; Deranlot, C.; Hamzić, A.; Broto, J.M.; et al. High Mobility in $LaAlO_3/SrTiO_3$ Heterostructures: Origin, Dimensionality, and Perspectives. *Phys. Rev. Lett.* **2007**, *98*, 216803. [CrossRef] [PubMed]
26. Yoshimatsu, K.; Yasuhara, R.; Kumigashira, H.; Oshima, M. Origin of metallic states at the heterointerface between the band insulators $LaAlO_3$ and $SrTiO_3$. *Phys. Rev. Lett.* **2008**, *101*, 026802. [CrossRef] [PubMed]
27. Zhang, L.; Zhou, X.F.; Wang, H.T.; Xu, J.J.; Li, J.; Wang, E.; Wei, S.H. Origin of insulating behavior of the p-type $LaAlO_3/SrTiO_3$ interface: Polarization-induced asymmetric distribution of oxygen vacancies. *Phys. Rev. B* **2010**, *82*, 125412. [CrossRef]
28. Reinle-Schmitt, M.; Cancellieri, C.; Li, D.; Fontaine, D.; Medarde, M.; Pomjakushina, E.; Schneider, C.; Gariglio, S.; Ghosez, P.; Triscone, J.M.; et al. Tunable Conductivity Threshold at Polar Oxide Interfaces. *Nat. Commun.* **2012**, *3*, 932. [CrossRef] [PubMed]
29. Liu, Z.; Li, C.; Lü, W.; Huang, X.; Huang, Z.; Zeng, S.; Qiu, X.; Huang, L.; Annadi, A.; Chen, J.; et al. Origin of the Two-Dimensional Electron Gas at $LaAlO_3/SrTiO_3$ Interfaces: The Role of Oxygen Vacancies and Electronic Reconstruction. *Phys. Rev. X* **2013**, *3*, 021010.
30. Kleinman, D.A. Theory of Second Harmonic Generation of Light. *Phys. Rev.* **1962**, *128*, 1761–1775. [CrossRef]
31. Lüpke, G. Characterization of semiconductor interfaces by second-harmonic generation. *Surf. Sci. Rep.* **1999**, *35*, 75–161. . [CrossRef]

32. Li, P.; Jiang, X.; Huang, M.; Kang, L.; Chen, S.; Gali, A.; Huang, B. Defect engineering of second-harmonic generation in nonlinear optical semiconductors. *Cell Rep. Phys. Sci.* **2022**, *3*, 101111. [CrossRef]
33. Shen, Y. Surfaces probed by nonlinear optics. *Surf. Sci.* **1994**, *299–300*, 551–562. [CrossRef]
34. Paparo, D.; Rubano, A.; Marrucci, L. Optical second-harmonic generation selection rules and resonances in buried oxide interfaces: The case of $LaAlO_3/SrTiO_3$. *J. Opt. Soc. Am. B* **2013**, *30*, 2452–2460. [CrossRef]
35. Rubano, A.; Fiebig, M.; Paparo, D.; Marino, A.; Maccariello, D.; di Uccio, U.S.; Granozio, F.M.; Marrucci, L.; Richter, C.; Paetel, S.; et al. Spectral and spatial distribution of polarization at the $LaAlO_3/SrTiO_3$ interface. *Phys. Rev. B* **2011**, *83*, 155405. [CrossRef]
36. Harris, D.; Bertolucci, M. *Symmetry and Spectroscopy: An Introduction to Vibrational and Electronic Spectroscopy*, new ed.; Dover Publications: Mineola, NY, USA, 1989.
37. Rubano, A.; Scigaj, M.; Sánchez, F.; Herranz, G.; Paparo, D. Optical second harmonic generation from $LaAlO_3/SrTiO_3$ interfaces with different in-plane anisotropies. *J. Phys. Cond. Matter* **2019**, *32*, 135001. [CrossRef] [PubMed]
38. Rubano, A.; Aruta, C.; di Uccio, U.S.; Granozio, G.M.; Marrucci, L.; Günter, T.; Fink, T.; Fiebig, M.; Paparo, D. Electronic states at polar/nonpolar interfaces grown on $SrTiO_3$ studied by optical second harmonic generation. *Phys. Rev. B* **2013**, *88*, 245434. [CrossRef]
39. Rubano, A.; Günter, T.; Fink, T.; Paparo, D.; Marrucci, L.; Cancellieri, C.; Gariglio, S.; Triscone, J.M.; Fiebig, M. Influence of atomic termination on the $LaAlO_3/SrTiO_3$ interfacial polar rearrangement. *Phys. Rev. B* **2013**, *88*, 035405. [CrossRef]
40. Cardona, M. Optical Properties and Band Structure of $SrTiO_3$ and $BaTiO_3$. *Phys. Rev.* **1965**, *140*, A651. [CrossRef]
41. Trepakov, V.; Dejneka, A.; Markovin, P.; Lynnyk, A.; Jastrabik, L. A 'soft electronic band' and the negative thermooptic effect in strontium titanate. *New J. Phys.* **2009**, *11*, 083024. [CrossRef]
42. Savoia, A.; Paparo, D.; Perna, P.; Ristic, Z.; Salluzzo, M.; Granozio, F.M.; di Uccio, U.S.; Richter, C.; Thiel, S.; Mannhart, J.; et al. Polar catastrophe and electronic reconstructions at the $LaAlO_3/SrTiO_3$ interface: Evidence from optical second harmonic generation. *Phys. Rev. B* **2009**, *80*, 075110. [CrossRef]
43. Ogawa, N.; Miyano, K.; Hosoda, M.; Higuchi, T.; Bell, C.; Hikita, Y.; Hwang, H. Enhanced lattice polarization in $SrTiO_3/LaAlO_3$ superlattices measured using optical second-harmonic generation. *Phys. Rev. B* **2009**, *80*, 081106. [CrossRef]
44. Santander-Syro, A.; Copie, O.; Kondo, T.; Fortuna, F.; Pailhes, S.; Weht, R.; Qiu, X.; Bertran, F.; Nicolaou, A.; Taleb-Ibrahimi, A.; et al. Two-dimensional electron gas with universal subbands at the surface of $SrTiO_3$. *Nature* **2011**, *469*, 189–193. [CrossRef]
45. Cantoni, C.; Gazquez, J.; Granozio, F.M.; Oxley, M.; Varela, M.; Lupini, A.; Pennycook, S.; Aruta, C.; di Uccio, U.S.; Perna, P.; et al. Electron Transfer and Ionic Displacements at the Origin of the 2D Electron Gas at the LAO/STO Interface: Direct Measurements with Atomic-Column Spatial Resolution. *Adv. Mater.* **2012**, *24*, 3952–3957. [CrossRef] [PubMed]
46. Luca, G.D.; Rubano, A.; Gennaro, E.D.; Khare, A.; Granozio, F.M.; di Uccio, U.S.; Marrucci, L.; Paparo, D. Potential-well depth at amorphous-$LaAlO_3$/crystalline-$SrTiO_3$ interfaces measured by optical second harmonic generation. *App. Phys. Lett.* **2014**, *104*, 1603.
47. Warusawithana, M.; Richter, C.; Mundy, J.; Roy, P.; Ludwig, J.; Paetel, S.; Heeg, T.; Pawlicki, A.; Kourkoutis, L.; Zheng, M.; et al. $LaAlO_3$ stoichiometry is key to electron liquid formation at $LaAlO_3/SrTiO_3$ interfaces. *Nat. Commun.* **2013**, *4*, 2351. [CrossRef] [PubMed]
48. Rubano, A.; Luca, G.D.; Schubert, J.; Wang, Z.; Zhu, S.; Schlom, D.; Marrucci, L.; Paparo, D. Polar asymmetry of $La_{(1-\delta)}Al_{(1+\delta)}O_3/SrTiO_3$ heterostructures probed by optical second harmonic generation. *App. Phys. Lett.* **2015**, *107*, 101603. [CrossRef]
49. Annadi, A.; Zhang, Q.; Wang, X.R.; Tuzla, N.; Gopinadhan, K.; Lu, W.; Barman, A.R.; Liu, Z.; Srivastava, A.; Saha, S.; et al. Anisotropic two-dimensional electron gas at the $LaAlO_3/SrTiO_3$ (110) interface. *Nat. Commun.* **2013**, *4*, 1838. [CrossRef]
50. Pentcheva, R.; Pickett, W. Charge localization or itineracy at $LaAlO_3/SrTiO_3$ interfaces: Hole polarons, oxygen vacancies, and mobile electrons. *Phys. Rev. B* **2006**, *74*, 035112. [CrossRef]
51. Popović, Z.; Satpathy, S.; Martin, R. Origin of the Two-Dimensional Electron Gas Carrier Density at the $LaAlO_3/SrTiO_3$ Interface. *Phys. Rev. Lett.* **2008**, *101*, 256801. [CrossRef]
52. Nanda, B.; Satpathy, S. Electronic phases and phase separation in the Hubbard-Holstein model of a polar interface. *Phys. Rev. B* **2011**, *83*, 195114. [CrossRef]
53. Ariando; Wang, X.; Baskaran, G.; Liu, Z.; Huijben, J.; Yi, J.; Annadi, A.; Barman, A.R.; Rusydi, A.; Dhar, S.; et al. Electronic phase separation at the $LaAlO_3/SrTiO_3$ interface. *Nat. Commun.* **2011**, *2*, 188. [CrossRef]
54. Kalabukhov, A.; Boikov, Y.; Serenkov, I.; Sakharov, V.; Popok, V.; Gunnarsson, R.; Börjesson, J.; Ljustina, N.; Olsson, E.; Winkler, D.; et al. Cationic Disorder and Phase Segregation in $LaAlO_3/SrTiO_3$ Heterointerfaces Evidenced by Medium-Energy Ion Spectroscopy. *Phys. Rev. Lett.* **2009**, *103*, 146101. [CrossRef]
55. Günter, T.; Rubano, A.; Paparo, D.; Lilienblum, M.; Marrucci, L.; Granozio, F.M.; di Uccio, U.S.; Jany, R.; Richter, C.; Mannhart, J.; et al. Spatial inhomogeneities at the $LaAlO_3/SrTiO_3$ interface: Evidence from second harmonic generation. *Phys. Rev. B* **2012**, *86*, 235418. [CrossRef]
56. Rubano, A.; Günter, T.; Lilienblum, M.; Aruta, C.; Granozio, F.M.; di Uccio, U.S.; Marrucci, L.; Paparo, D.; Fiebig, M. Optical second harmonic imaging as a diagnostic tool for monitoring epitaxial oxide thin-film growth. *App. Surf. Sci.* **2015**, *327*, 413–417. [CrossRef]
57. Nordlander, J.; Luca, G.D.; Strkalj, N.; Fiebig, M.; Trassin, M. Probing Ferroic States in Oxide Thin Films Using Optical Second Harmonic Generation. *Appl. Sci.* **2018**, *8*, 570. [CrossRef]

58. Rubano, A.; Günter, T.; Fiebig, M.; Granozio, F.M.; Marrucci, L.; Paparo, D. Ultrafast modification of the polarity at $LaAlO_3/SrTiO_3$ interfaces. *Phys. Rev. B* **2018**, *97*, 035438. [CrossRef]
59. Yamada, Y.; Sato, H.; Hikita, Y.; Hwang, H.; Kanemitsu, Y. Measurement of the Femtosecond Optical Absorption of $LaAlO_3/SrTiO_3$ Heterostructures: Evidence for an Extremely Slow Electron Relaxation at the Interface. *Phys. Rev. Lett.* **2013**, *111*, 047403. [CrossRef] [PubMed]
60. Cancellieri, C.; Mishchenko, A.; Aschauer, U.; Filippetti, A.; Faber, C.; Barišić, O.; Rogalev, V.; Schmitt, T.; Nagaosa, N.; Strocov, V. Polaronic metal state at the $LaAlO_3/SrTiO_3$ interface. *Nat. Commun.* **2016**, *7*, 10386. [CrossRef]

materials

MDPI

Article

Atomic Layer Deposition of Titanium Oxide-Based Films for Semiconductor Applications—Effects of Precursor and Operating Conditions

Vladyslav Matkivskyi [1,*], Oskari Leiviskä [2], Sigurd Wenner [3], Hanchen Liu [2] Ville Vähänissi [2], Hele Savin [2], Marisa Di Sabatino [1] and Gabriella Tranell [1]

[1] Department of Materials Science and Engineering, Norwegian University of Science and Technology, Alfred Getz vei 2B, 7034 Trondheim, Norway; marisa.di.sabatino@ntnu.no (M.D.S.); gabriella.tranell@ntnu.no (G.T.)
[2] Department of Electronics and Nanoengineering, Aalto University, Tietotie 3, 02150 Espoo, Finland; oskari.leiviska@aalto.fi (O.L.); hanchen.liu@aalto.fi (H.L.); ville.vahanissi@aalto.fi (V.V.); hele.savin@aalto.fi (H.S.)
[3] SINTEF Industry, Høgskoleringen 5, 7034 Trondheim, Norway; sigurd.wenner@ntnu.no
* Correspondence: vladyslav.matkivskyi@ntnu.no

Abstract: Two widely used atomic layer deposition precursors, Tetrakis (dimethylamido) titanium (TDMA-Ti) and titanium tetrachloride ($TiCl_4$), were investigated for use in the deposition of TiOx-based thin films as a passivating contact material for solar cells. This study revealed that both precursors are suited to similar deposition temperatures (150 °C). Post-deposition annealing plays a major role in optimising the titanium oxide (TiO_x) film passivation properties, improving minority carrier lifetime (τ_{eff}) by more than 200 µs. Aluminium oxide deposited together with titanium oxide (AlO_y/TiO_x) reduced the sheet resistance by 40% compared with pure TiO_x. It was also revealed that the passivation quality of the (AlO_y/TiO_x) stack depends on the precursor and ratio of AlO_y to TiO_x deposition cycles.

Keywords: atomic layer deposition; titanium oxide; passivation films; ALD precursors

Citation: Matkivskyi, V.; Leiviskä, O.; Wenner, S.; Liu, H.; Vähänissi, V.; Savin, H.; Di Sabatino, M.; Tranell, G. Atomic Layer Deposition of Titanium Oxide-Based Films for Semiconductor Applications—Effects of Precursor and Operating Conditions. *Materials* **2023**, *16*, 5522. https://doi.org/10.3390/ma16165522

Academic Editors: Antonio Polimeni, Sebastiano Vasi and Ulderico Wanderlingh

Received: 21 June 2023
Revised: 1 August 2023
Accepted: 6 August 2023
Published: 8 August 2023

1. Introduction

Titanium oxide materials (TiO_x) are used in a wide range of applications such as batteries [1], medicine [2], semiconductors [3], and solar photovoltaic (PV) cells [4,5]. Among the many metal oxides, titanium oxide has the closest band offset with c-Si [6], making it suitable for semiconductor/PV applications. In addition, the high thermal stability [6,7] and availability of deposition/formation methods [8–10] for TiO_x make its processing favourable.

In solar cell fabrication, titanium oxide first appeared as part of the anti-reflective coating [5] and is still used in protective coatings for solar cells [11]. Currently, the carrier selectiveness and passivation properties of TiO_x have gained interest in the semiconductor industry [12,13]. This is mainly due to the rapid market implementation of the TOPcon (passivated contact cell) [14] architecture, which has brought forward potential new passivation materials such as TiO_x.

As a crystalline material, titanium dioxide has three different crystal polymorphs: rutile [15], brookite [16], and anatase [15]. Anatase and rutile are the most common in TiO_2 fabrication, and the anatase phase is the most desirable for semiconductor applications because of its conducting and passivation properties [17]. The semiconductor industry has developed a range of techniques for TiO_x deposition such as atomic layer deposition (ALD) [18], E-beam electron evaporation [19], magnetron sputtering [20], and chemical vapour deposition [21], and the choice of deposition process affects the final quality of the TiO_x film [10].

ALD of titanium oxide has evolved as one of the main alternatives to sputtering, which is widely applied in the industry [22] and is under constant development to obtain highly passivating TiO_x films [23]. Both Tetrakis (dimethylamido) titanium (TDMA-Ti) and titanium tetrachloride ($TiCl_4$) precursors for TiO_x ALD are proven to provide high-quality TiO_x passivation layers for silicon solar cells [24–26]. The deposition of passivating TiO_x from the $TiCl_4$ precursor was presented in the work of Yu et al. [25]. The minimum achieved surface recombination velocity was 44.24 cm/s for deposition at 200° C. The deposition temperature of TiO_x plays a crucial role, as documented by previous studies [27,28]. In 2021, Liu et al. achieved a high-performance passivating electron contact by deposition of aluminium oxide (Al_2O_3) and TiO_x in a stack using ALD at 150 °C [24]. With the application of TDMA-Ti and H_2O precursors, a high minority carrier lifetime (τ_{eff}) of 1.9 ms was obtained with a low contact resistivity of 0.1 $\Omega \cdot cm^2$ [24]. Most ALD systems allow precise control of parameters such as deposition temperature [29], temperature of precursors [30], purging time [31], and number of deposition cycles. However, uniform control of the gases inside the reactor is not possible as the precursor gas distribution is controlled by the carrier gas. As such, large-scale industrial implementation of TiO_x ALD with a consistent product outcome is often challenging.

To approach consistent results, key parameters such as carrier gas flow, precursor pulse duration, purge duration, and deposition temperature must be optimised. In the current work, we present a comparison between two different precursors, namely TDMA-Ti and $TiCl_4$, for ALD of TiO_x using different deposition and post-annealing process conditions. Furthermore, the effect of introducing aluminium oxide (AlO_y) in the stack with TiO_x was investigated as aluminium oxide is widely applied not only as a passivation [23] layer but also as a tandem layer with other metal oxides [23,32] to improve the electronic properties of metal oxides such as resistance [33]. Electronic and crystalline properties of the deposited TiO_x and TiO_x/AlO_y layers obtained in the current work were analysed using a range of techniques such as microwave photo-conductance decay (μ-PCD), four-point sheet resistance probe, and transmission electron microscopy (TEM).

2. Experimental Materials and Procedures

2.1. Material Preparation

Experimental samples were prepared using laser scribing of as-cut (100), n-type wafers into 3 × 3 cm size. The initial thickness and resistivity of the wafers were 180 μm and 1–3.5 $\Omega \cdot cm$, respectively.

An HNA solution ($1HNO_3$ (75%):$1CH_5COOH$ (99.7%):0.2HF(45%)) was used for surface damage removal. Following damage removal, the samples were cleaned in an RCA 2 (0.1HCl (37%):$0.2H_2O_2$ (30%):H_2O) solution at 70 °C. Next, the samples were immersed in a low-concentration HF solution for the removal of native oxide. The last part of the sample preparation was cleaning the samples in an RCA 1 ($0.2NH_4OH$ (30%):H_2O_2:H_2O) solution at 70–75 °C (while forming so-called "RCA oxide").

2.2. ALD Deposition

Atomic layer deposition of titanium oxide in this work was conducted from two different precursors TDMA-Ti and $TiCl_4$. Two different ALD systems were used: Beneq TFS-500 for deposition of TiO_x using a $TiCl_4$ precursor (at Aalto University, Helsinki, Finland) and Savannah S100 with TDMA-Ti as a precursor (at the Norwegian University of Science and Technology, Trondheim, Norway). In both cases, the second precursor was water. Each deposition set consisted of six samples, which, after the deposition, were split into three parallels of two samples. Post-deposition annealing (PDA) was performed in a rapid thermal annealing system (RTP Allwin). The second part of this study consisted of stack layer deposition of aluminium oxide (AlO_y) and TiO_x. Deposition of the AlOy in the AlO_y/TiO_x stack was conducted using Trimethylaluminium (TMA) precursor as the first precursor and water as the second (on both ALD equipment Beneq TFS-500 and

Savannah S100). The AlO_y:TiO_x deposition ratios were: 1:1, 1:5, 1:30, and 1:60, respectively. All experimental details are presented in Table 1.

Table 1. Main experimental conditions.

Deposited Thickness of TiOx (nm)	3, 6, 9,12, 15, 20
Deposition temperature of TiOx from TDMA-Ti (°C)	120, 140, 150, 180, 200, 210
Post deposition annealing temperature (°C)	200, 250, 300, 350, 400
Deposition temperature of TiO_x from $TiCl_4$ precursor (°C)	120, 150, 180, 210
Deposition layer stack ration AlO_y:TIO_x	1:1, 1:5, 1:30, 1:60

Post-deposition annealing of the samples with the RTP system was conducted according to the RTP profile demonstrated in Figure 1 with the plateau temperatures outlined in Table 1.

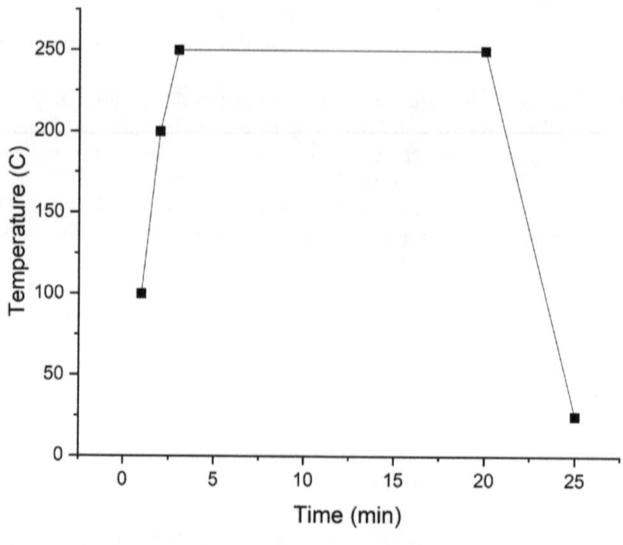

Figure 1. RTP system post-deposition annealing profile with a plateau of 250 °C.

2.3. Characterisation

After deposition and post-deposition annealing, the passivation and material properties of the thin films were studied. Minority carrier lifetime (MCLT) was measured using the transient photo-conductance decay (PCD) method with a Sinton WCT-120 tool, and the sheet resistance of the samples was determined using a CMT-SR2000N four-point probe. Five points per sample were measured using the four-point probe, and median values were calculated. Three samples per deposition condition such as thickness and deposition temperature were analysed. The samples with the highest MCLT and sheet resistance for both precursors were studied using TEM (transmission electron microscopy). Preliminary film thickness measurements were carried out using spectroscopic ellipsometry (Woollam M2000) for the TDMA-Ti precursor deposition. The thickness of the deposited TiO_x films was also measured using TEM. The TEM analysis was performed with a Helios 5 plasma-focused ion beam (PFIB). Electron-deposited carbon and subsequently platinum were used as protection layers. Cross-section lift-out and thinning were performed using Xe ions, finishing with 2 kV ions. TEM/annular dark field scanning TEM (ADF-STEM) was performed using a double-corrected JEOL ARM-200F cold field emission microscope at 200 kV. Energy-dispersive X-ray spectroscopy (EDS) and electron energy loss spectroscopy (EELS) were performed using JEOL Centurio and GIF Quantum detectors. Scanning precession electron diffraction was performed with a JEOL 2100F microscope at 200 kV.

3. Results and Discussion

3.1. TiO_x Films

The first part of this study concentrated on the deposition conditions for the two different TiO_x precursors. For both precursors, titanium oxide was deposited in the temperature range from 120 °C to 200 °C. During the experiments, an intended 15 nm thick TiO_x layer was deposited at these temperatures with subsequent μ-PCD measurements. Initial ellipsometry measurements showed a 1.5 nm average thickness error for the TDMA-Ti-deposited TiO_x films for the temperature range 120–200 °C (see Supplementary Material for details). Thus, the growth per cycle rate (GPC) was assumed to be constant for the investigated temperature range, in accordance with the previously reported value of 0.5 Å/cycle [34]. Figure 2 presents the measured values of the minority carrier lifetime for the as-deposited samples obtained from the $TiCl_4$ and TDMA-Ti precursors, respectively. As seen in Figure 2, the TiO_x films deposited at 150 °C have the highest measured MCLT for both precursors and thus we conclude, in accordance with previous studies, that this temperature is most efficient in promoting the formation of the TiO_x anatase phase [35].

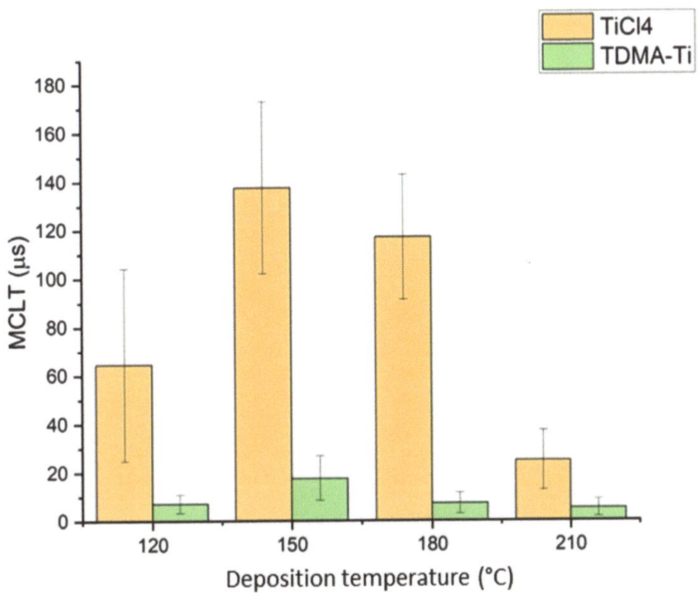

Figure 2. As-deposited median values of measured MCLT for an intended 15 nm thick TiO_x film deposited at different temperatures. Error bars illustrate the standard deviation for each condition/precursor.

The $TiCl_4$ precursor films consistently display a higher minority carrier lifetime than the TDMA-Ti precursor. All MCLT results are, however, much lower than those presented in other studies [23,24]. This might be caused by the deposition recipe setup parameters or precursor distribution in the reactor. In some studies, the only reported results are obtained after post-deposition annealing, which has proven successful in increasing MCLT [24,26]. Thus, the samples deposited at 150 °C were annealed at 200 °C–400 °C for 20 min in N_2 atmosphere with subsequent MCLT measurements. It was demonstrated that annealing at 250 °C significantly improves surface passivation for both precursors (Figure 3). The increase in MCLT for the annealed TiO_x deposited from the TDMA-Ti is larger than that of $TiCl_4$, making the effect of the precursor on the MCLT after annealing insignificant at temperatures above 200 °C. Such a difference in the post-annealing MCLT improvement might be related to initial differences in structure, density, or thickness of the as-deposited films which, following annealing, become less significant. A more detailed structural analysis of the as-deposited films may provide additional information on the morphological differences between the as-deposited films from $TiCl_4$ and TDMA-Ti. Additionally, we can see that annealing at or above 300 °C does not positively affect MCLT. Finally, it is possible that at temperatures over 250 °C, crystal nucleation is initiated at multiple sites in the oxide film, which results in phase transformation of TiO_x at these sites.

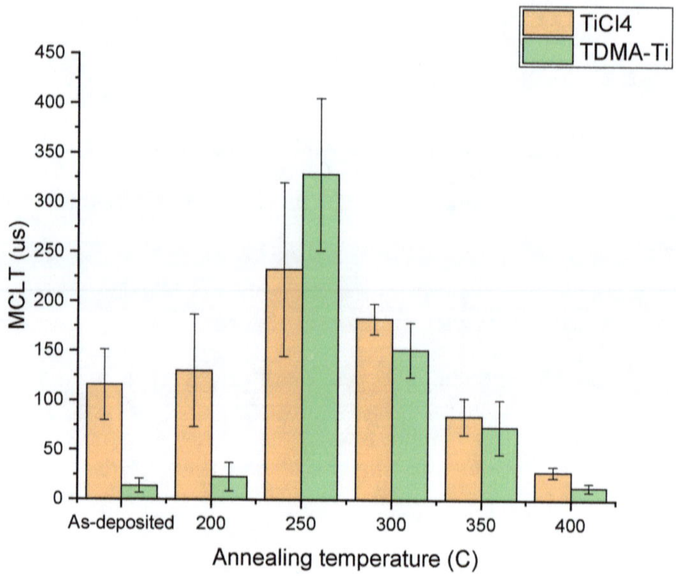

Figure 3. Measured median MCLT of the deposited 9 nm TiOx films deposited at 150 °C, post-annealed at different temperatures. Error bars illustrate the standard deviation for each condition/precursor.

The thickness of the deposited layer is an important characteristic of the passivation- and carrier-selective layers. As such, TMOs (transparent metal oxides) typically display a clear correlation between the thickness of the deposited oxide and the passivation efficiency [36]. Thus, the relationship between thickness and passivation efficiency, as well as thickness and sheet resistance, were investigated for both precursors. TiO_x was deposited with an intended thickness range of 3–20 nm at 150 °C, with subsequent annealing. MCLT and sheet resistance were measured both before and after annealing.

Figure 4 presents the measured MCLT as a function of the targeted TiO_x thickness for TMDA-Ti and $TiCl_4$ precursors for the samples deposited at 150 °C and annealed at 250 °C and for the non-annealed samples. In both cases, the passivation efficiency increased with TiO_x layer thickness up to 9 nm. Above 9 nm thickness, the passivation efficiency decreased for the TDMA-Ti precursor. The annealing treatment did not provide a significant improvement for 12, 15, or 20 nm of deposited TiO_x (TDMA-Ti). The reason for such poor response might be related to the thickness of the films. In the case of the reaction in the TiO_x layer or the reaction between the TiO_x layer and the Si/SiO_y interface, for the thicker oxide layers, a longer annealing time may be required. Thus, further experiments correlating the post-deposition annealing time with layer thickness may be needed.

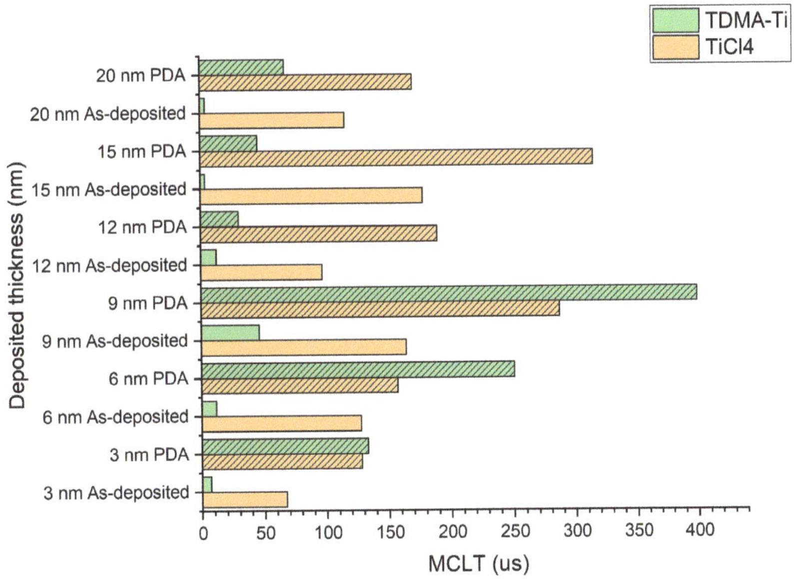

Figure 4. Measured median MCLT for the different deposited thicknesses of TiO_x before and after PDA.

The sheet resistance of the TiO_x layers deposited at 150 °C and annealed at 250 °C was measured using the four-probe method. Figure 5a presents the measured sheet resistance for films deposited with both precursors at different thicknesses of the titanium oxide. We can observe a clear trend in resistance increase with thickness for the $TiCl_4$ precursor while the resistance of the layers deposited from TDMA-Ti remains in the range of 170–250 Ω/sq. For the 20 nm thickness TiO_x deposited with the $TiCl_4$ precursor, some measured resistance values are clear outliers (Figure 5b). Such abnormal values increase the median resistance and standard deviation. The low resistance of the TDMA-Ti (TiO_x) might be due to a higher resistance-change threshold of the deposited TiOx caused by a change in the current path through the material. The TDMA-Ti-deposited TiO_x film resistance may increase faster after a certain thickness higher than 20 nm, as in other Ti-based materials [37,38]. However, to prove this theory, additional experimental work is required.

3.2. Al_2O_3/TiO_x Stack Films

Techniques using aluminium oxide in a stack with titanium oxide [39] or aluminium-doped TiO_x [24] are under development, mostly because of the improved conductivity of such layers in comparison to TiO_x alone. Such passivation layers are one of the alternatives to ultra-thin a-Si:H passivation [40]. In this work, AlO_y/TiO_x stacks were studied as a possible alternative to mono-TiO_x layers, quantifying the potential improvement for each precursor.

Figure 6 illustrates the effect of introducing aluminium oxide in the TiO_x stack (intended 9 nm) for both precursors. For neither precursor, there is clear improvement in MCLT when introducing AlO_y in the stack.

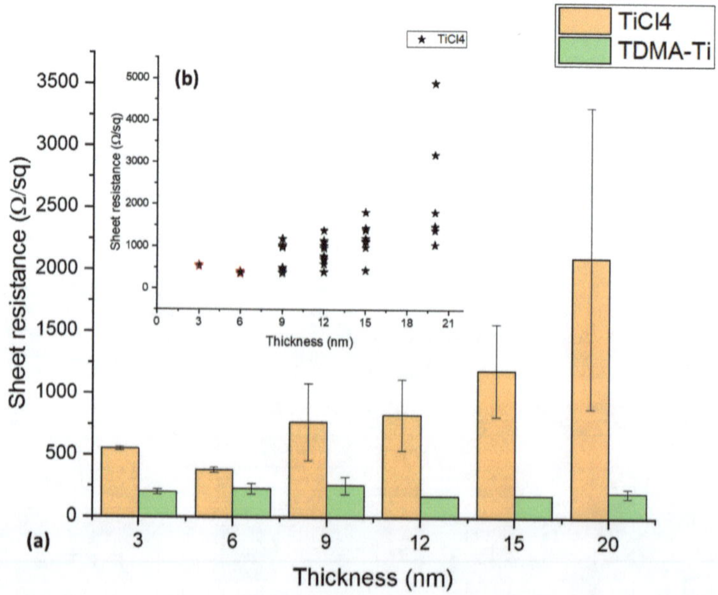

Figure 5. (**a**) Median sheet resistance measurement of the different thicknesses for the deposited TiO_x films after PDA. (**b**) Resistance measurement data distribution for the $TiCl_4$ precursor. Error bars illustrate the standard deviations for each condition/precursor.

Following the MCLT analysis of the samples, the sheet resistance of the Al_2O_3/TiO_x stacks was measured. In Figure 7, no clear differences in resistance between the precursor stacks were observed. However, resistance was reduced by more than 40% (770 Ω/sq to 405 Ω/sq) from the original values of the TiO_x layer for the $TiCl_4$ precursor. The measured sheet resistance of the deposited AlO_y/TiO_x stacks was at approximately the same level for all AlO_y/TiO_x ratios. The measured sheet resistances are detailed in the Supplementary Materials.

3.3. *TEM Analysis of Deposited Films*

In order to better understand the differences observed between precursors, a TEM analysis of three samples for each of the two titanium oxide ALD precursors was carried out. These included the targeted 9 nm of pure TiO_x annealed and as-deposited, along with the targeted 9 nm deposited AlO_y/TiO_x stacks.

First the achieved, as opposed to targeted, TiO_x and AlO_y/TiO_x layer thicknesses were measured. The actual deposited thickness of the targeted 9 nm layers for the $TiCl_4$ precursor was 16.6 nm for the annealed AlO_y/TiOx (1:60) stack (Figure 8a), while for the annealed TiO_x (Figure 8b) and as-deposited TiO_x (Figure 8c), the thicknesses were 8 nm and 7.7 nm, respectively. For the TDMA-Ti precursor, the thickness was 12.1 nm for the annealed AlO_y/TiO_x (1:1) stack (Figure 8d), 10.1 nm for the as-deposited TiO_x (Figure 8f), and 10.1 nm for the annealed TiO_x (Figure 8e). As such, the intended deposited thicknesses and the actual thickness of the layers were slightly different for the single TiO_x deposited films. The thickness of the AlO_y/TiOx (1:60) stack using the $TiCl_4$ precursor was 7.6 nm thicker than expected, while the TDMA-Ti precursor thickness of the AlO_y/TiO_x (1:1) stack was 3.1 nm thicker than intended. However, the pure TiO_x layers deposited using the $TiCl_4$ precursors are somewhat thinner than intended, while those deposited using TDMA-Ti are slightly thicker than expected.

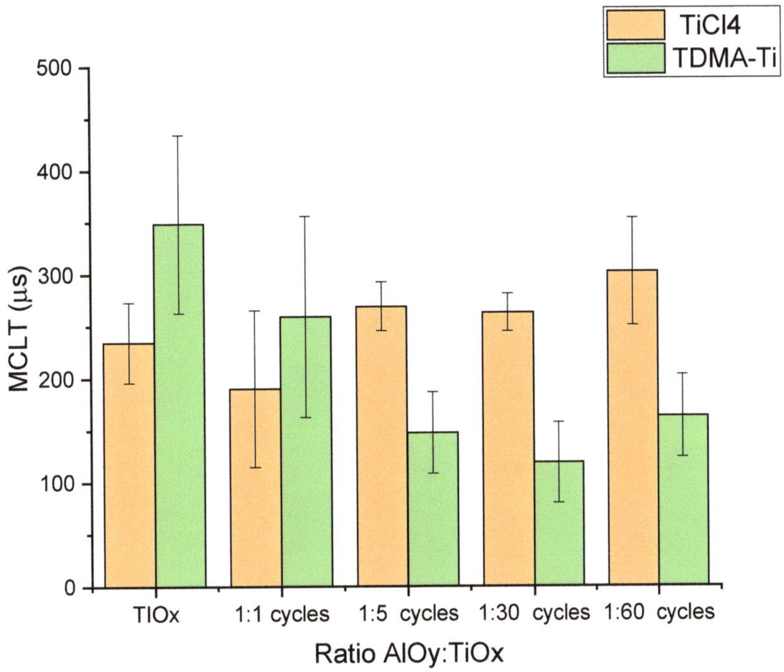

Figure 6. Measured median MCLT of the deposited AlO_y/TiO_x stack depending on the AlO_y:TiOx deposition cycle ratio after annealing at 300 °C. Error bars illustrate the standard deviations for each condition/precursor.

As the initial ellipsometry measurements showed small thickness deviations for the TDMA-Ti-deposited TiO_x films, these measurements were not continued for the bulk of samples. Hence, there may be differences in the thicknesses measured using ellipsometry and those measured using TEM. Further work to compare the thicknesses obtained using TEM and ellipsometry should hence be performed in future studies.

The TEM-measured thicknesses of the AlO_y/TiO_x stacks also deviate from those intended. During the deposition, precursors were pulsed into the reactor one by one after a certain amount of time (purging time). It is possible that the purging time for AlO_y was not long enough to remove the products of the TMA and H_2O precursor reaction [41], which resulted in the additional growth of the AlO_y layer during the next pulses of water precursor into the reactor. Thus, during the deposition of TiO_x and AlO_y, additional TMA precursors may remain in the reactor, resulting in the additional growth of the oxide layer.

Elemental mapping using a combination of electron energy loss spectroscopy (EELS) and energy-dispersive X-ray spectroscopy (EDS) was also carried out for each of the samples (Figure 9). While EDS is not a quantitatively reliable tool, it gives a good indication of the relative concentration of elements. A comparison of the AlO_y/TiO_x-deposited stacks revealed that Al is distributed across the whole oxide layer for both precursors, while the deposition process was performed layer-by-layer. In the case of the 1:1 AlO_y/TiO_x ratio, such an Al distribution might be possible due to the proposed TDMA-Ti precursor residue in the reactor. In the case of the 1:60 ratio, where only 1 deposition cycle of AlO_y was performed for 60 deposition cycles of the TiO_x, and each cycle had its own purging, the probability of the AlO_y being distributed across the whole oxide layer caused by TDMA-Ti precursor residues is low. However, the thickness of the deposited layer per cycle is approximately 0.05 nm for TiO_x, while it is 0.1 nm for the AlO_y [34]. As a result, for the 1:1 cycle deposition, the thin film will consist of 0.05–0.1 nm thick layers, which are not possible to identify with the resolution of the TEM instrument used in this work. Thus, the conclusion is that the EDS analysis does not have a high enough resolution to give an accurate composition of the individual deposited layers.

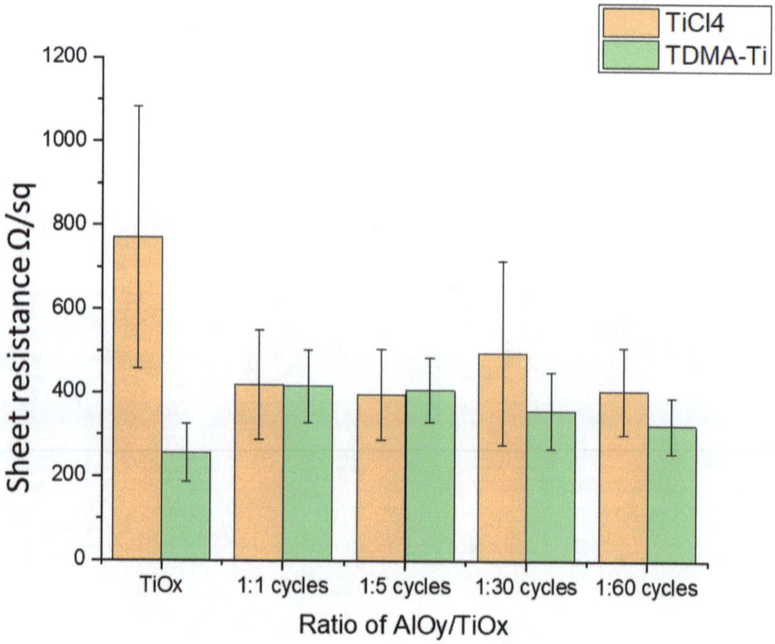

Figure 7. Measured median sheet resistance of the deposited AlO_y/TiO_x stack depending on the AlO_y:TiOx deposition cycle ratio after annealing at 300 °C. Error bars illustrate the standard deviation for each condition/precursor.

For both the as-deposited and annealed TiO_x layers, a difference in the Ti:O ratio between precursors was found. For both precursors, there was also a difference between the as-deposited and annealed TiO_x, which corresponded with the layer thickness differences. The composition of the TiO_x films from elemental mapping measured in atomic percentage using EDS analysis is presented in Table 2. Although the values obtained using EDS cannot be claimed to be quantitatively exact, relative differences are more reliable.

Table 2. Atomic percentage of the TEM EDS-analysed samples. Atomic percentage error indicates summed EDS fitting error during a measurement.

	Ti, Atomic %	O, Atomic %
As-deposited TiO_x ($TiCl_4$)	22.83 ± 0.25	77.17 ± 0.78
As-deposited TiO_x (TDMA-Ti)	29.53 ± 0.21	70.47 ± 0.47
Annealed TiO_x ($TiCl_4$)	20.76 ± 0.31	79.24 ± 0.99
Annealed TiO_x (TDMA-Ti)	26.09 ± 0.20	73.80 ± 0.48

As summarised in Table 2 and suggested by the MCLT measurement results, it is possible that the oxygen to titanium content dictates passivation properties. We can see that the annealed samples of both precursors display a higher oxygen-to-titanium ratio and a higher minority carrier lifetime than the as-deposited samples. This phenomenon was also observed in previous work for the E-beam evaporated TiO_x layer [42], indicating that a higher oxygen content improves the passivation properties of the TiO_x layer.

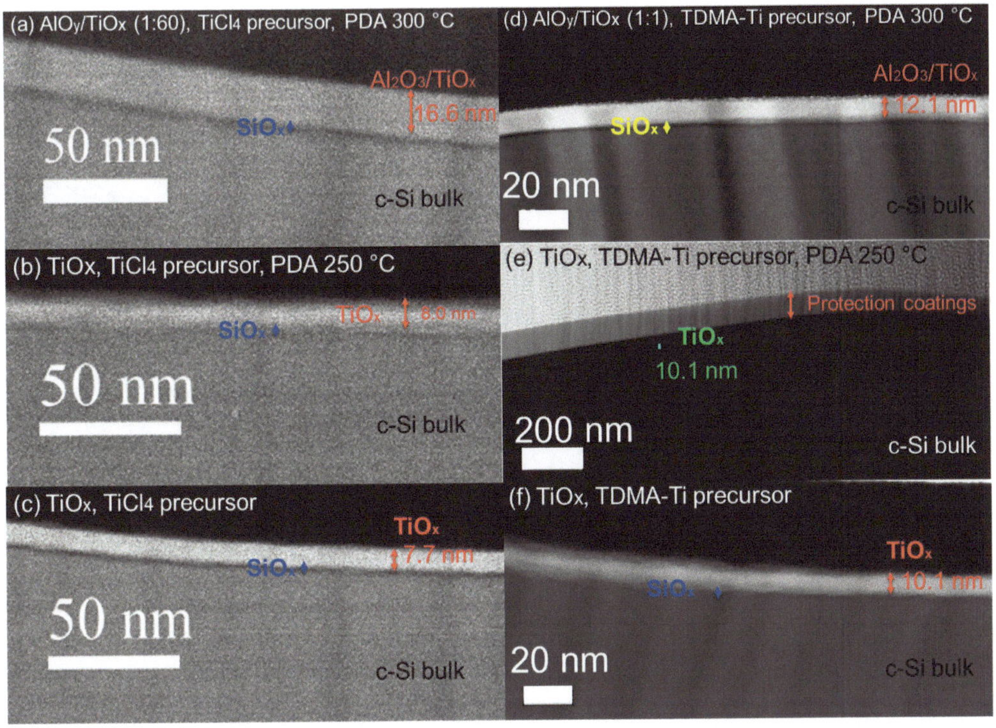

Figure 8. ADF-STEM image showing the TiOx and AlO_y/TiO_x stack layers: (**a**) Annealed at 300 °C AlO_y/TiO_x stack (1:60), $TiCl_4$ precursor, (**b**) annealed at 250 °C TiO_x layer, $TiCl_4$ precursor, (**c**) TiO_x as-deposited, $TiCl_4$ precursor, (**d**) annealed at 300 °C AlO_y/TiO_x stack (1:1), TDMA-Ti precursor, (**e**) annealed at 250 °C TiO_x layer, TDMA-Ti precursor, and (**f**) TiO_x as-deposited, TDMA-Ti precursor.

The TEM work revealed that the TDMA-Ti-deposited TiO_x was 2 nm thicker than the $TiCl_4$-based film. Such a thickness difference might affect the MCLT, as passivation tends to increase with the thickness of the oxide layer [43,44]. The same trend is presented in this work for the annealed TiO_x from 3 to 9 nm thickness (Figure 2). This theory would also explain the higher MCLT of the TDMA-Ti-deposited TiO_x layer (at thicknesses $\leq$ 9 nm intended thickness) over the $TiCl_4$ precursor. However, if we assume that each deposited TiO_x TDMA-Ti thickness is 2 nm thicker than intended, in accordance with Figure 8, the 9 nm thick layer of TDMA-Ti TiO_x still has a relatively higher MCLT than the 12 nm thick $TiCl_4$-based TiO_x film.

During the high-resolution TEM analysis of the annealed samples obtained from the $TiCl_4$ precursor, crystallised areas were discovered. Figure 10 shows a high-resolution TEM image of annealed TiO_x ($TiCl_4$) film (8 nm) with faint signals of crystallinity. For neither of the AlO_y/TiO_x stacks ($TiCl_4$) nor the as-deposited TiO_x ($TiCl_4$) samples, lattice diffraction was indicated. For the annealed TDMA-Ti precursor samples, the same signals of crystallinity were also revealed with electron diffraction of the selected area. Figure 11 shows the diffraction patterns in the annealed TiO_x for both precursors. Like the $TiCl_4$ precursor, the as-deposited samples from TDMA-Ti showed no traces of crystallinity.

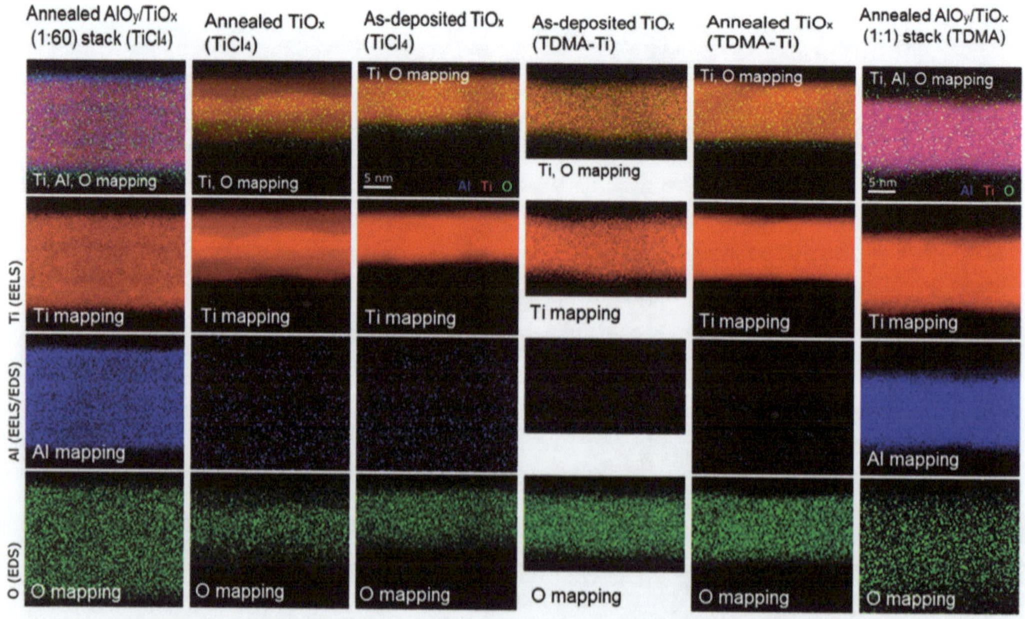

Figure 9. EDS/EELS elemental mapping of the samples analysed with TEM.

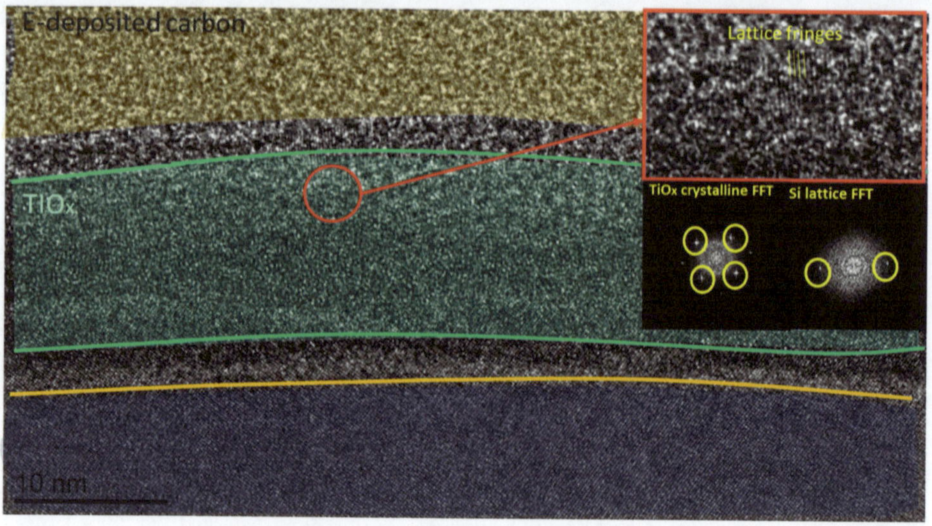

Figure 10. High-resolution TEM image showing the annealed TiO_x sample ($TiCl_4$ precursor) with Fourier transforms (FFTs) of the layer attached (FFT zone axis (001)).

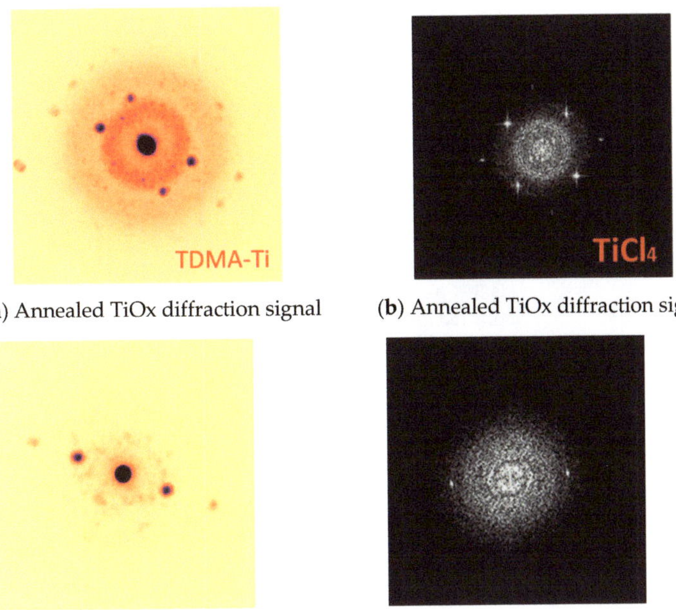

(**a**) Annealed TiOx diffraction signal (**b**) Annealed TiOx diffraction signal

(**c**) Si lattice diffraction signal (**d**) Si lattice diffraction signal

Figure 11. Diffraction peaks from the selected area electron diffraction scanning for TiO_x deposited with different precursors.

Signs of nano-crystallinity for annealed TiO_x have been reported in previous studies. Phase-change from amorphous TiO_x to anatase in atomic-layer-deposited TiO_x (from titanium IV isopropoxide TTIP precursor) was discovered in a post-deposition annealing process at 450 °C [45]. In our case, the temperature of the PDA process was lower at 250 °C, and the crystallinity signal was weak. However, this may indicate that already at 250 °C, the phase-change process is initiated in ultra-thin TiO_x films. In order to gain valuable information on the crystallinity of TiO_x films, conducting grazing incidence X-ray diffraction analysis must be considered in future research.

Crystallinity might affect passivation quality. During the TiO_x phase change process, additional oxygen might be absorbed, which will decrease oxygen deficiency in the film. Another possible reason for the improvement in the measured MCLT after the PDA process is the additional growth of the silicon oxide (SiO_2) at the Si/TiO_x interface [42,46]. Previous studies have presented opposite effects on passivation quality with the crystallisation of TiO_x [46]. However, a previous study by the current first author on E-beam-deposited TiO_x indicated the same passivation improvement with film crystallisation [42].

Neither of the annealed AlO_y/TiO_x films deposited from $TiCl_4$ and TDMA-Ti precursors showed signs of crystallinity despite a higher annealing temperature (300 °C). These samples also showed less improvement in passivation quality after the PDA process. This might be related to a too-low temperature for "activation" of AlO_y passivation [47] in conjunction with the absence of a real stack, as aluminium is distributed across the whole film. This probably affected the sheet resistance of the mixed oxide layer in comparison with TiO_x alone, as Al is widely used as a dopant to improve the conductivity of MO films [24,48]. The aluminium oxide presence might also hinder the crystal formation (phase change) of TiO_x. The reason for such an effect might be the formation of a new aluminium/titanium compound (Al_x-O_y-Ti_z) in the passivation layer, resulting in the distribution of Al across the oxide layer. Thus, a different temperature is required to improve the passivation quality of the layer as well as initiate a phase change process for TiO_x.

4. Conclusions

This work aimed at comparing titanium oxide precursors ($TiCl_4$ and TDMA-Ti) for atomic layer deposition of passivating TiO_x. The precursors were compared with respect to the deposition temperature, post-deposition annealing composition, and properties of deposited films.

A deposition temperature of 150 °C was found to be optimal for both precursors. It was confirmed that post-deposition annealing of the deposited TiO_x improves the passivation quality of the oxides. The optimal post-deposition annealing temperature of 250 °C was also similar for the two precursors.

The maximum achieved minority carrier lifetime (τ_{eff}) for the deposited 9 nm of TiO_x was 398 μs for the TDMA-Ti precursor and 286 μs for $TiCl_4$. A TEM analysis of the as-deposited and annealed TiOx revealed signals of nano-crystallinity for the post-deposition annealed TiO_x films for both precursors, indicating that a phase change of the atomic-layer-deposited TiO_x possibly starts at 250 °C. However, there is no evidence of a correlation between crystallinity and passivation improvement.

Aluminium oxide (AlO_y)/titanium oxide (TiO_x) stacks showed higher passivation quality than single TiO_x films for both precursors with a maximum (τ_{eff}) of 311 μs for the $TiCl_4$ precursor. For the $TiCl_4$ precursor, the passivation quality increased with a decreasing AlO_y:TiO_x cycle ratio (from 1:1 to 1:60), while for the TDMA-Ti precursor, it was vice versa (from 1:60 to 1:1). Furthermore, sheet resistance measurements showed that the AlO_y/TiO_x stack had much lower resistance than the pure TiO_x layer. Thus, we conclude that the Al presence in TiO_x films decreases sheet resistance, while additional experimental work, such as optimising annealing temperature, is required to improve the passivation quality of the AlO_y/TiO_x stack.

5. Future Perspectives

In order to continue the work of the current study, some future perspectives may be presented: The performance of the deposition process for the two different ALD equipment units should be tuned to the same GPC for the two precursors and, if possible, also ensure the same precursor distribution during the deposition. This might be helpful to better evaluate the correlation between deposited thickness and differences in sample characteristics (e.g., MCLT). Moreover, further post-deposition analysis of the obtained films such as grazing incidence X-ray diffraction analysis and transfer length measurements (TLM) will give a better understanding of film conductivity and crystallinity. Further experimental work investigating the effect of post-annealing duration on different TiO_x film thicknesses is recommended.

Supplementary Materials: The following supporting information can be downloaded at: https://www.mdpi.com/article/10.3390/ma16165522/s1.

Author Contributions: Methodology, V.M., O.L., H.L., V.V., M.D.S. and G.T.; Formal analysis, V.M.; Investigation, V.M., O.L. and S.W.; Resources, H.S.; Writing—original draft, V.M.; Writing—review & editing, V.M. and G.T.; Supervision, H.S., M.D.S. and G.T. All authors have read and agreed to the published version of the manuscript.

Funding: The FIB preparation was performed at the SMART-H infrastructure, funded by The Research Council of Norway (RCN), grant 296197. The TEM work was carried out at the NORTEM infrastructure, RCN grant 197405.

Institutional Review Board Statement: Not applicable.

Informed Consent Statement: Not applicable.

Data Availability Statement: Available on request to authors.

Acknowledgments: The authors gratefully acknowledge financial support from the Research Council of Norway through the research project "DiamApp" (project no. 280831). The authors acknowledge the provision of facilities and technical support by Micronova Nanofabrication Centre in Espoo, Finland within the OtaNano research infrastructure at Aalto University.

Conflicts of Interest: The authors declare no conflict of interest.

References

1. Han, X.; Ouyang, M.; Lu, L.; Li, J. Cycle Life of Commercial Lithium-Ion Batteries with Lithium Titanium Oxide Anodes in Electric Vehicles. *Energies* **2014**, *7*, 4895–4909. [CrossRef]
2. Jackson, M.J.; Kopac, J.; Balazic, M.; Bombac, D.; Brojan, M.; Kosel, F. Titanium and Titanium Alloy Applications in Medicine. In *Surgical Tools and Medical Devices*; Ahmed, W., Jackson, M.J., Eds.; Springer International Publishing: Cham, Switzerland, 2016; pp. 475–517. [CrossRef]
3. Li, W.; Fu, T.; Xie, F.; Yu, S.; He, S. The multi-staged formation process of titanium oxide nanotubes and its thermal stability. *Mater. Lett.* **2007**, *61*, 730–735. [CrossRef]
4. Cui, J.; Allen, T.; Wan, Y.; Mckeon, J.; Samundsett, C.; Yan, D.; Zhang, X.; Cui, Y.; Chen, Y.; Verlinden, P.; et al. Titanium oxide: A re-emerging optical and passivating material for silicon solar cells. *Sol. Energy Mater. Sol. Cells* **2016**, *158*, 115–121. [CrossRef]
5. Afzal, A.; Habib, A.; Ulhasan, I.; Shahid, M.; Rehman, A. Antireflective Self-Cleaning TiO2 Coatings for Solar Energy Harvesting Applications. *Front. Mater.* **2021**, *8*, 687059. [CrossRef]
6. Melskens, J.; van de Loo, B.W.H.; Macco, B.; Black, L.E.; Smit, S.; Kessels, W.M.M. Passivating Contacts for Crystalline Silicon Solar Cells: From Concepts and Materials to Prospects. *IEEE J. Photovoltaics* **2018**, *8*, 373–388. [CrossRef]
7. Richards, B.S. *Novel Uses of Titanium Dioxide for Silicon Solar Cells*; UNSW Sydney: Kensington, Australia, 2002.
8. Horprathum, M.; Chindaudom, P.; Limsuwan, P. A Spectroscopic Ellipsometry Study of TiO_2 Thin Films Prepared by dc Reactive Magnetron Sputtering: Annealing Temperature Effect. *Chin. Phys. Lett.* **2007**, *24*, 1505–1508. [CrossRef]
9. Eiamchai, P.; Chindaudom, P.; Pokaipisit, A.; Limsuwan, P. A spectroscopic ellipsometry study of TiO2 thin films prepared by ion-assisted electron-beam evaporation. *Curr. Appl. Phys.* **2009**, *9*, 707–712. [CrossRef]
10. Hitosugi, T.; Yamada, N.; Nakao, S.; Hirose, Y.; Hasegawa, T. Properties of TiO2-based transparent conducting oxides. *Phys. Status Solidi (a)* **2010**, *207*, 1529–1537. [CrossRef]
11. Mazumder, M.K.; Sharma, R.; Biris, A.S.; Zhang, J.; Calle, C.; Zahn, M. Self-Cleaning Transparent Dust Shields for Protecting Solar Panels and Other Devices. *Part. Sci. Technol.* **2007**, *25*, 5–20. [CrossRef]
12. Kang, X.; Liu, S.; Dai, Z.; He, Y.; Song, X.; Tan, Z. Titanium Dioxide: From Engineering to Applications. *Catalysts* **2019**, *9*, 191. [CrossRef]
13. Nowotny, J. *Oxide Semiconductors for Solar Energy Conversion: Titanium Dioxide*; CRC Press: Boca Raton, FL, USA, 2011.
14. Zeng, Y.; Tong, H.; Quan, C.; Cai, L.; Yang, Z.; Chen, K.; Yuan, Z.; Wu, C.-H.; Yan, B.; Gao, P.; et al. Theoretical exploration towards high-efficiency tunnel oxide passivated carrier-selective contacts (TOPCon) solar cells. *Sol. Energy* **2017**, *155*, 654–660. [CrossRef]
15. Kamisaka, H.; Adachi, T.; Yamashita, K. Theoretical study of the structure and optical properties of carbon-doped rutile and anatase titanium oxides. *J. Chem. Phys.* **2005**, *123*, 084704. [CrossRef] [PubMed]
16. Murakami, N.; Kamai, T.-A.; Tsubota, T.; Ohno, T. Novel hydrothermal preparation of pure brookite-type titanium(IV) oxide nanocrystal under strong acidic conditions. *Catal. Commun.* **2009**, *10*, 963–966. [CrossRef]
17. Bakri, A.S.; Sahdan, M.Z.; Adriyanto, F.; Raship, N.A.; Said, N.D.M.; Abdullah, S.A.; Rahim, M.S. Effect of annealing temperature of titanium dioxide thin films on structural and electrical properties. *AIP Conf. Proc.* **2017**, *1788*, 030030. [CrossRef]
18. Kukli, K.; Ritala, M.; Schuisky, M.; Leskelä, M.; Sajavaara, T.; Keinonen, J.; Uustare, T.; Hårsta, A. Atomic layer deposition of titanium oxide from TiI_4 and H_2O_2. *Chem. Vap. Depos.* **2000**, *6*, 303–310. [CrossRef]
19. Jang, H.K.; Whangbo, S.W.; Choi, Y.K.; Chung, Y.D.; Jeong, K.; Whang, C.N.; Lee, Y.S.; Lee, H.-S.; Choi, J.Y.; Kim, G.H.; et al. Titanium oxide films on Si(100) deposited by e-beam evaporation. *J. Vac. Sci. Technol. A* **2000**, *18*, 2932–2936. [CrossRef]
20. César, R.; Barros, A.; Doi, I.; Diniz, J.; Swart, J. Thin titanium oxide films obtained by RTP and by sputtering. In Proceedings of the 2014 29th Symposium on Microelectronics Technology and Devices (SBMicro), Aracaju, Brazil, 1–5 September 2014; pp. 1–4.
21. Mathur, S.; Kuhn, P. CVD of titanium oxide coatings: Comparative evaluation of thermal and plasma assisted processes. *Surf. Coatings Technol.* **2006**, *201*, 807–814. [CrossRef]
22. Meng, L.-J.; dos Santos, M. Investigations of titanium oxide films deposited by d.c. reactive magnetron sputtering in different sputtering pressures. *Thin Solid Films* **1993**, *226*, 22–29. [CrossRef]
23. Liao, B.; Hoex, B.; Aberle, A.G.; Chi, D.; Bhatia, C.S. Excellent c-Si surface passivation by low-temperature atomic layer deposited titanium oxide. *Appl. Phys. Lett.* **2014**, *104*, 253903. [CrossRef]
24. Liu, Y.; Sang, B.; Hossain, A.; Gao, K.; Cheng, H.; Song, X.; Zhong, S.; Shi, L.; Shen, W.; Hoex, B.; et al. A novel passivating electron contact for high-performance silicon solar cells by ALD Al-doped TiO_2. *Sol. Energy* **2021**, *228*, 531–539. [CrossRef]
25. Yu, I.-S.; Chang, I.-H.; Cheng, H.-E.; Lin, Y.-S. Surface passivation of c-Si by atomic layer deposition TiO_2 thin films deposited at low temperature. In Proceedings of the 2014 IEEE 40th Photovoltaic Specialist Conference (PVSC), Denver, CO, USA, 8–13 June 2014; pp. 1271–1274. [CrossRef]

26. Yu, I.-S.; Wang, Y.-W.; Cheng, H.-E.; Yang, Z.-P.; Lin, C.-T. Surface Passivation and Antireflection Behavior of ALD on n-Type Silicon for Solar Cells. *Int. J. Photoenergy* **2013**, *2013*, 431614. [CrossRef]
27. Hsu, C.-H.; Chen, K.-T.; Huang, P.-H.; Wu, W.-Y.; Zhang, X.-Y.; Wang, C.; Liang, L.-S.; Gao, P.; Qiu, Y.; Lien, S.-Y.; et al. Effect of Annealing Temperature on Spatial Atomic Layer Deposited Titanium Oxide and Its Application in Perovskite Solar Cells. *Nanomaterials* **2020**, *10*, 1322. [CrossRef]
28. Peng, T.; Xiao, X.; Ren, F.; Xu, J.; Zhou, X.; Mei, F.; Jiang, C. Influence of annealing temperature on the properties of TiO2 films annealed by ex situ and in situ TEM. *J. Wuhan Univ. Technol. Sci. Ed.* **2012**, *27*, 1014–1019. [CrossRef]
29. Parsons, G.N.; George, S.M.; Knez, M. Progress and future directions for atomic layer deposition and ALD-based chemistry. *MRS Bull.* **2011**, *36*, 865–871. [CrossRef]
30. Xie, Q.; Jiang, Y.-L.; Detavernier, C.; Deduytsche, D.; Van Meirhaeghe, R.L.; Ru, G.-P.; Li, B.-Z.; Qu, X.-P. Atomic layer deposition of TiO2 from tetrakis-dimethyl-amido titanium or Ti isopropoxide precursors and H_2O. *J. Appl. Phys.* **2007**, *102*, 083521. [CrossRef]
31. Park, H.K.; Yang, B.S.; Park, S.; Kim, M.S.; Shin, J.C.; Heo, J. Purge-time-dependent growth of ZnO thin films by atomic layer deposition. *J. Alloys Compd.* **2014**, *605*, 124–130. [CrossRef]
32. Saarenpää, H.; Niemi, T.; Tukiainen, A.; Lemmetyinen, H.; Tkachenko, N. Aluminum doped zinc oxide films grown by atomic layer deposition for organic photovoltaic devices. *Sol. Energy Mater. Sol. Cells* **2010**, *94*, 1379–1383. [CrossRef]
33. Merisalu, J.; Arroval, T.; Kasikov, A.; Kozlova, J.; Rähn, M.; Ritslaid, P.; Aarik, J.; Tamm, A.; Kukli, K. Engineering of atomic layer deposition process for titanium-aluminum-oxide based resistively switching medium. *Mater. Sci. Eng. B* **2022**, *282*. [CrossRef]
34. Peron, M.; Cogo, S.; Bjelland, M.; Bin Afif, A.; Dadlani, A.; Greggio, E.; Berto, F.; Torgersen, J. On the evaluation of ALD TiO2, ZrO2 and HfO2 coatings on corrosion and cytotoxicity performances. *J. Magnes. Alloy.* **2021**, *9*, 1806–1819. [CrossRef]
35. Hanaor, D.A.H.; Sorrell, C.C. Review of the anatase to rutile phase transformation. *J. Mater. Sci.* **2010**, *46*, 855–874. [CrossRef]
36. Macco, B.; Deligiannis, D.; Smit, S.; van Swaaij, R.A.C.M.M.; Zeman, M.; Kessels, W.M.M. Influence of transparent conductive oxides on passivation of a-Si:H/c-Si heterojunctions as studied by atomic layer deposited Al-doped ZnO. *Semicond. Sci. Technol.* **2014**, *29*, 122001. [CrossRef]
37. Musschoot, J.; Xie, Q.; Deduytsche, D.; Berghe, S.V.D.; Van Meirhaeghe, R.; Detavernier, C. Atomic layer deposition of titanium nitride from TDMAT precursor. *Microelectron. Eng.* **2009**, *86*, 72–77. [CrossRef]
38. Fox, G.R.; Potrepka, D.M.; Polcawich, R.G. Dependence of {111}-textured Pt electrode properties on TiO2 seed layers formed by thermal oxidation. *J. Mater.Sci. Mater. Electron.* **2018**, *29*, 412–426. [CrossRef]
39. Temperton, R.H.; Gibson, A.; O'Shea, J.N. In situ XPS analysis of the atomic layer deposition of aluminium oxide on titanium dioxide. *Phys. Chem. Chem. Phys.* **2018**, *21*, 1393–1398. [CrossRef]
40. Cheng, X.; Marstein, E.S.; Haug, H.; Di Sabatino, M. Double Layers of Ultrathin a-Si:H and SiNx for Surface Passivation of n-type Crystalline Si Wafers. *Energy Procedia* **2016**, *92*, 347–352. [CrossRef]
41. Weckman, T.; Laasonen, K. First principles study of the atomic layer deposition of alumina by TMA–H 2 O-process. *Phys. Chem. Chem. Phys.* **2015**, *17*, 17322–17334. [CrossRef]
42. Matkivskyi, V.; Lee, Y.; Seo, H.S.; Lee, D.-K.; Park, J.-K.; Kim, I. Electronic-beam evaporation processed titanium oxide as an electron selective contact for silicon solar cells. *Curr. Appl. Phys.* **2021**, *32*, 98–105. [CrossRef]
43. Rafique, R.; Tonny, K.N.; Sharmin, A.; Mahmood, Z.H. Study on the Effect of Varying Film Thickness on the Transparent Conductive Nature of Aluminum Doped Zinc Oxide Deposited by Dip Coating. *Mater. Focus* **2018**, *7*, 707–713. [CrossRef]
44. Kale, A.S.; Nemeth, W.; Harvey, S.P.; Page, M.; Young, D.L.; Agarwal, S.; Stradins, P. Effect of silicon oxide thickness on polysilicon based passivated contacts for high-efficiency crystalline silicon solar cells. *Sol. Energy Mater. Sol. Cells* **2018**, *185*, 270–276. [CrossRef]
45. Hussin, R.; Choy, K.L.; Hou, X.H. Growth of TiO2 Thin Films by Atomic Layer Deposition (ALD). *Adv. Mater. Res.* **2016**, *1133*, 352–356. [CrossRef]
46. Chen, X.Y.; Lu, Y.F.; Tang, L.J.; Wu, Y.H.; Cho, B.J.; Xu, X.J.; Dong, J.R.; Song, W.D. Annealing and oxidation of silicon oxide films prepared by plasma-enhanced chemical vapor deposition. *J. Appl. Phys.* **2005**, *97*, 014913. [CrossRef]
47. Dingemans, G.; Kessels, W. Status and prospects of Al2O3-based surface passivation schemes for silicon solar cells. *J. Vac. Sci. Technol. A Vac.Surf. Film.* **2012**, *30*, 040802. [CrossRef]
48. Kelly, P.; Zhou, Y.; Postill, A. A novel technique for the deposition of aluminium-doped zinc oxide films. *Thin Solid Films* **2003**, *426*, 111–116. [CrossRef]

materials

MDPI

Article

Asymmetric TMO–Metal–TMO Structure for Enhanced Efficiency and Long-Term Stability of Si-Based Heterojunction Solar Cells

Yoon-Chae Jung [1], Young-Jin Yu [1], Yu-Kyung Kim [1], Jin Hee Lee [2], Jung Hwa Seo [3] and Jea-Young Choi [4,*]

1 Department of Metallurgical Engineering, Dong-A University, Busan 49315, Republic of Korea
2 Department of Chemical Engineering (BK21 Four Graduate Program), Dong-A University, Busan 49315, Republic of Korea
3 Department of Physics, University of Seoul, Seoul 02504, Republic of Korea
4 Department of Materials Sciences & Engineering, Dong-A University, Busan 49315, Republic of Korea
* Correspondence: cjy4395@dau.ac.kr

Abstract: In this study, we fabricated Si-based heterojunction solar cells (HSCs) with an asymmetric TMO–metal–TMO (TMT) structure using both MoO_3 and V_2O_5 as the hole-selective contacts. Our HSCs offer enhanced long-term stability and effective passivation for crystal defects on the Si sur-face. We analyzed the oxygen vacancy state and surface morphology of the MoO_3- and V_2O_5-TMO thin films using X-ray photoelectron spectroscopy and atomic force microscopy to investigate their passivation characteristics for Si surface defects. From the measured minority carrier lifetime, V_2O_5 revealed a highly improved lifetime (590 μs) compared to that of MoO_3 (122.3 μs). In addition, we evaluated the long-term stability of each TMO thin film to improve the operation stability of the HSCs. We deposited different types of TMOs as the top- and bottom-TMO layers and assessed the effect of the thickness of each TMO layer. The fabricated asymmetric TMT/Si HSCs showed noticeable improvements in efficiency (7.57%) compared to 6.29% for the conventional symmetric structure which used the same TMO material for both the top and bottom layers. Furthermore, in terms of long-term stability, the asymmetric TMT/Si HSCs demonstrated an efficiency that was 250% higher than that of symmetric TMT/Si HSCs, as determined via power conversion efficiency degradation over 2000 h which is mainly attributed by the lower oxygen vacancy of the top-TMO, V_2O_5. These results suggest that the asymmetric TMT structure is a promising approach for the fabrication of low-cost and high-efficiency Si-based HSCs with enhanced long-term stability.

Keywords: transition metal oxide; asymmetric; passivation; long-term stability; silicon; heterojunction solar cell

Citation: Jung, Y.-C.; Yu, Y.-J.; Kim, Y.-K.; Lee, J.H.; Seo, J.H.; Choi, J.-Y. Asymmetric TMO–Metal–TMO Structure for Enhanced Efficiency and Long-Term Stability of Si-Based Heterojunction Solar Cells. *Materials* **2023**, *16*, 5550. https://doi.org/10.3390/ma16165550

Academic Editors: Sebastiano Vasi and Ulderico Wanderlingh

Received: 4 July 2023
Revised: 14 July 2023
Accepted: 3 August 2023
Published: 9 August 2023

1. Introduction

In recent decades, crystalline silicon (c-Si) has become a prominent technology in the global photovoltaic market due to its numerous advantages, including high efficiency, long-term stability, abundant and non-toxic silicon material, and a favorable energy bandgap that allows for effective light absorption in the visible region [1]. c-Si solar cells can be classified into homojunction and heterojunction solar cells based on the type of junction material. The p–n junction-based homojunction c-Si solar cell is favored for its high power conversion efficiency (PCE) and stability [2]. However, the production of homojunction solar cells is costly, as it requires a complex manufacturing process that inevitably includes high-temperature doping processes, ≥800 °C [1,3]. To address the limitations of conventional homojunction c-Si solar cells, researchers have been exploring the use of heterojunction structures to reduce manufacturing costs and improve PCE. Generally, heterojunctions can be fabricated through a simpler and less costly process, and materials that allow for the formation of junctions at relatively low temperatures are attracting significant attention.

These materials include graphene [4], transparent conductive oxides (TCOs) [5,6], transition metal oxides (TMOs) [7,8], organic materials [9,10], and perovskite [11,12]. Among the materials mentioned, transition metal oxides (TMOs) have been shown to be effective carrier-selective materials that provide electrical properties and photovoltaic performance that are suitable for dopant-free Si-based heterojunction solar cells (HSCs) [13–15]. TMOs have a wide range of work functions, from 3 to 7 eV, which allows for the formation of both hole- and electron-selective contacts by selecting appropriate materials [16]. TMOs with high work functions close to 7 eV, such as molybdenum trioxide (MoO_3), vanadium pentoxide (V_2O_5), and tungsten trioxide (WO_3), are particularly suitable for hole-selective layers [17]. Additionally, the relatively large energy bandgap of TMOs ($E_g > 3$ eV) also minimizes parasitic absorption when deposited as front contacts in HSCs [16,18]. Moreover, the presence of TMOs on Si surfaces leads to the formation of a thermodynamically spontaneous silicon oxide (SiO_x, where $0 \leq x \leq 2$) layer at the TMO/Si interface, which contributes as a passivation layer for the Si surface [13,19]. TMOs provide two types of passivation features for the Si surface: (1) chemical passivation, which decreases the density of Si surface defects through chemical bonding between the oxygen in TMOs and the dangling bonds on the Si surface; and (2) field-effect passivation, which arises from the high work function of the TMO and produces an imbalance of electron and hole carrier concentrations at the TMO/Si interface [20,21]. These properties of TMOs as passivation layers and as carrier-selective contacts can reduce carrier recombination on the Si surface, increase the efficiency of charge carrier extraction, and thus enhance the PCE of HSCs [22,23].

Despite the advantages of TMO introduced HSCs, the development of high-efficiency TMO/Si HSCs faces several critical challenges originated from TMO layers, including the following: (1) poor passivation effect (i.e., high carrier recombination velocity); (2) limited long-term stability; and (3) high sheet resistance (R_{sheet}). First, the low passivation effect results in a reduced PCE because it decreases the minority carrier lifetime [15]. The formation of a low-quality SiO_x passivation layer during the deposition of TMOs on the Si surface is the root cause of this low passivation effect [24]. Second, the long-term stability of the solar cells is limited by the high oxygen vacancy density in TMOs formed during deposition by thermal evaporation, which reacts with airborne molecules (e.g., water) over time, degrading the TMO work function and ultimately the stability of the HSC [25]. Lastly, as a dielectric material, TMO thin film (TF) inherently exhibits a high R_{sheet} which inevitably hinders the extraction of photogenerated carriers from inside Si [26]. To achieve high-efficiency Si-based HSCs, it is crucial to develop methods that improve (1) the passivation performance for the Si surface; (2) the efficiency of charge carrier extraction; and (3) the long-term stability of the fabricated HSCs. The aim of this study is to simultaneously achieve improved efficiency and long-term stability of TMO/Si HSCs after introducing multilayered TMO with two different TMOs, namely, vanadium pentoxide (V_2O_5, VO) and molybdenum trioxide (MoO_3, MO). The optimal passivation characteristics of TMOs were attained through morphological and stoichiometric analysis of the two deposited TMO TFs (VO and MO), which were produced at low temperatures below 125 °C. However, it was believed that the single-layer TMO structure would limit the PCE of the HSC. Therefore, a TMO–Metal–TMO (TMT)-structured HSC was fabricated to further improve PCE. The TMT/Si HSC was fabricated by introducing an asymmetric TMT structure through thermal deposition. MO was used as the bottom-TMO (BTMO) due to its relatively low R_{sheet}, while VO was used as the top-TMO (TTMO) to enhance field-effect passivation. We confirmed that the VO TF used as a TTMO layer also offered enhanced long-term stability due to its low oxygen vacancy density. From the report, we successfully demonstrate that the introduction of an asymmetric TMT structure in a Si-based HSC significantly improves the PCE and long-term stability of the HSC, even with simple and low-temperature fabrication processes.

2. Experimental Section

2.1. Material and Sample Preparation

For Si, a double-sided polished n-type CZ Si wafer with a (100) orientation, thickness of 280 μm, and resistivity of 1.7–2.3 Ω·cm was employed. For sample preparation, the Si wafer was cut to the size of 2 × 2 cm for HSC fabrication, and it was cleaned through ultrasonication for 15 min with acetone, methanol, and distilled water (DI water) in the order. The wafer then underwent a standard RCA cleaning process ($NH_4OH : H_2O_2 : DI\ water = 1:1:5$) for 15 min to remove organic residues on the wafer, and the wafer was dipped for 1 min in 1% diluted HF solution to remove the native oxide. After the HF dip process, the VO and MO (Alfa Aesar, Haverhill, MA, USA, 99.995% powder) TFs were deposited using a molybdenum (Mo) boat via vacuum thermal evaporation. Each TMO was deposited at a rate of 0.2 Å/s and a vacuum level of 1×10^{-6} mbar at the target heating temperatures (no heating, 75 °C, and 125 °C). In the case of the metal electrode, Al and Ag were deposited to a thickness of 200 nm. at a rate of 1.0 Å/s and a vacuum level of 1×10^{-6} mbar. To examine the change in passivation characteristics with Si substrate temperature, HSC with a TMO/Si/TMO sandwich structure was fabricated by deposition after heating the substrate up to different target temperatures such as room temperature (RT, i.e., no-heating), 75, and 125 °C.

2.2. Characterization

The minority carrier lifetimes (τ_{eff}) of the Si samples with a single-TMO TF and a TMT multilayer were measured with a photoconductance decay system (WCT-120, Sinton Instrument Inc., Boulder, CO, USA). The surface morphology of the TMO TF was analyzed using atomic force microscopy (AFM, tapping mode, Multi-Mode V, Veeco, Oyster Bay, NY, USA). Furthermore, the chemical composition and oxygen deficiency of TMO TFs were analyzed using X-ray photoelectron spectroscopy (XPS, ESCALB-250XI, Thermo-Fischer Scientific, Waltham, MA, USA). The sheet resistance (R_{sheet}) of the TMO TF was measured using the transmission line method (TLM) after setting the voltage scan range −1.0 to 1.5 V. Finally, the PCE of the fabricated HSCs was investigated using a solar simulator under an air mass (AM) 1.5 G condition. (Note: to obtain the reliability of data, 3 to 5 identical samples were fabricated for each characterization.)

3. Results and Discussion

3.1. Effect of TMO on the Performance of TMO/Si HSCs

In this study, to evaluatie the Si surface passivation characteristics of two different TMOs according to the deposition temperatures, sandwich structured samples were fabricated with MO or VO deposited on both sides of the Si substrate at substrate temperatures of RT, 75, and 125 °C. The minority carrier lifetimes (τ_{eff}) of the samples were measured, and the results are presented in Figure 1a,b. In Figure 1b, for MO, the highest τ_{eff} (122.3 μs) was measured at RT, followed by a decrease in τ_{eff} with increasing substrate temperature. However, for VO, the highest τ_{eff} (590 μs) was measured at 75 °C. This substrate temperature-dependent τ_{eff} behavior is attributed to the change in the initial TMO/Si interface area (i.e., active region) for supplying oxygen atoms to the Si substrate as the aspect ratio (AR) of the initially deposited TMO nano island changes with increasing substrate temperature [24]. Figure 1d–i shows the AFM images of the nano islands according to the substrate temperature. Figure 1c reveals that the AR of the nano island is lowest at 75 °C and RT in the case of VO and MO, respectively, which is consistent with the τ_{eff} result shown in Figure 1b. Furthermore, in the case of VO, the lowest AR was observed at 75 °C, which confirms that the initially depositing VO can supply oxygen atoms to the Si surface more efficiently than the equivalent case in MO.

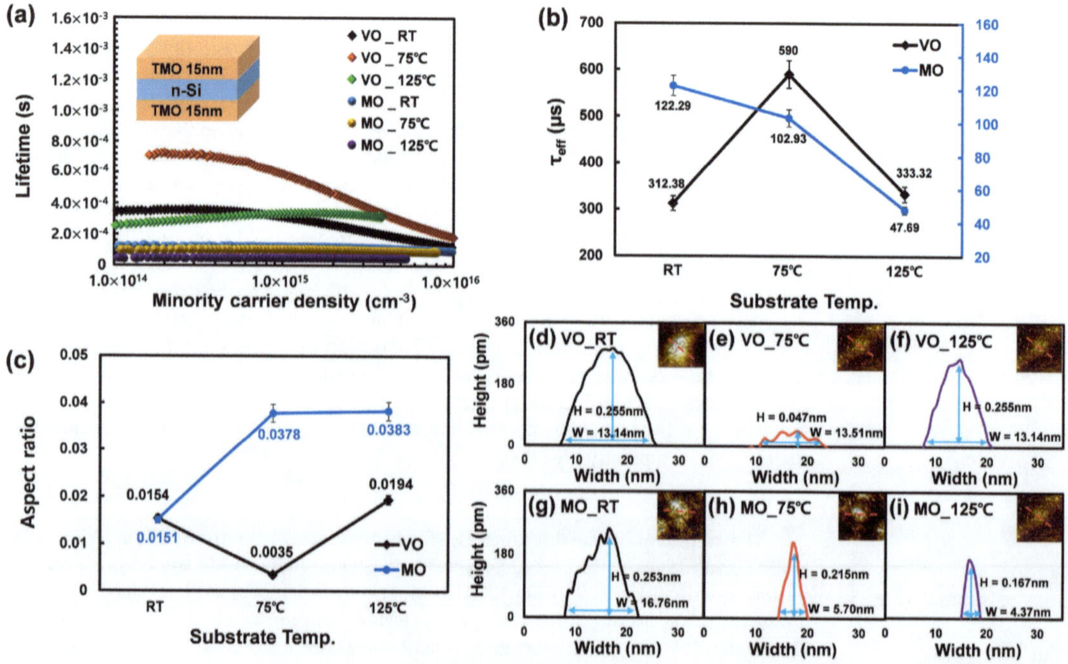

Figure 1. (**a**) τ_{eff} of two TMOs (VO and MO) deposited at different temperatures (RT, 75, and 125 °C); (**b**) extracted τ_{eff} values at the carrier injection level of 1.5×10^{15} cm^{-3}; (**c**) aspect ratio of TMO islands; (**d–i**) the width and height of each TMO island deposited at various substrate temperatures.

However, despite the low AR of the VO nano island, the Gibbs formation energy (ΔG) for the formation of SiO_2 from MO ($\Delta G^{MO \rightarrow SiO_2} = -406$ kJ/mol) is more negative than that from VO ($\Delta G^{VO \rightarrow SiO_2} = -285$ kJ/mol) [21]. Therefore, even though the formation of SiO_2 from MO is a thermodynamically more favorable, it is believed that the enhanced passivation performance of VO is due to factors other than SiO_x formation—specifically, the improved field-effect passivation of VO. The characteristics of the enhanced field-effect passivation of VO will be further discussed in Section 3.2. The performances of VO/Si and MO/Si HSCs were evaluated by depositing VO and MO on the Si surface, respectively, to analyze the relationship between the passivation characteristics of the TMO TFs and the PCE of the fabricated HSCs. Figure 2 shows the current density-voltage (J–V) curves, while Table 1 provides details on the solar cell performance parameters of the TMO/Si HSCs under both dark and illuminated conditions. Despite the exceptional Si surface passivation performance of VO as demonstrated in Figure 1b, the actual PCE of the MO/Si HSC was slightly higher than that of the VO/Si HSC, as shown in Table 1.

The performance parameters of the solar cells listed in Table 1 indicate that despite the better passivation performance of VO, it does not translate into an improvement in the actual PCE of the VO/Si HSC, which can most probably be attributed to the high series resistance (R_s) of the VO/Si HSC. This demonstrates that the VO material itself has a high parasitic resistance in the device compared to MO. The R_{sheet} of each TMO TF was determined through transmission line method (TLM) measurements, as shown in Figure 3 [27]. By comparing the R_{sheet} of each TMO TF, it was found that VO has a R_{sheet} that is 20% higher than that of MO followed by relatively lower charge extraction efficiency with VO TF. The high R_{sheet} of the VO TF is caused by its lower density of oxygen vacancies compared to that of the MO. To further examine these relationships, X-ray photoelectron spectroscopy (XPS) spectra were analyzed for each TMO TF.

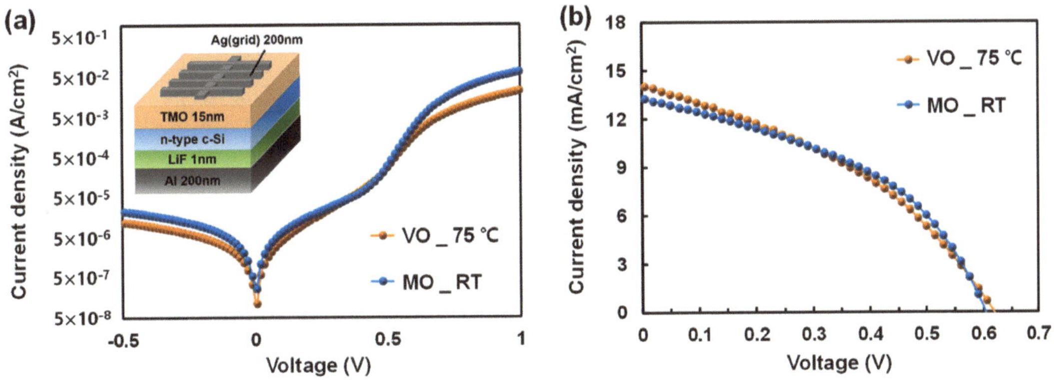

Figure 2. Current density–voltage (J–V) curves of TMO/Si HSCs under (**a**) dark and (**b**) light (AM1.5) conditions.

Table 1. Solar cell performance parameters of TMO/Si HSCs under light (AM1.5) conditions.

Sample	J_{sc} (mA/cm^2)	V_{oc} (mV)	FF (%)	R_{sh} ($\Omega\cdot cm^2$)	R_s ($\Omega\cdot cm^2$)	PCE (%)
VO_75 °C	14.09	618	37.81	14,102	5.65	3.29 ± 0.09
MO_RT	13.29	603	43.20	13,291	3.54	3.46 ± 0.11

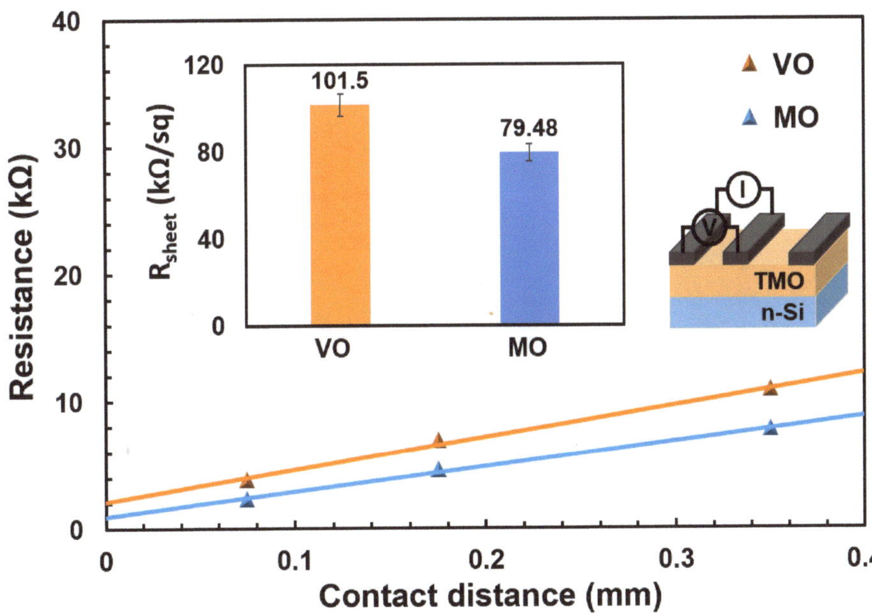

Figure 3. Sheet resistance of TMO TFs extracted from the ohmic current–voltage responses from TLM measurements (inset: structure of the fabricated TLM sample).

The XPS spectra of V 2p and Mo 3d are shown in Figure 4, and the integrated peak area ratio of the oxidation states of the two TMOs is presented in Table 2. In the case of VO, the V^{5+} ratio, which forms complete stoichiometric V_2O_5 in the deposited VO, was as high as 88.3%. However, for MO, the Mo^{6+} ratio, which forms MoO_3, was 80.5%, which is lower than the oxidation state ratio of VO. The difference in these values indicates that more metallic TFs were formed with MO compared to VO, which is believed to be the reason for the higher R_{sheet} of VO TF. The comparison of the resistance and passivation characteristics of the respective TMO TF, as well as the PCE of TMO/Si HSCs, confirm that the implementation of low resistance TMO is critical for high-efficiency TMO/Si HSCs. However, in terms of the TMO/Si HSC efficiency, considering the low PCE value of 3.46% despite the comparatively low R_s, further improvement in PCE should be required. In addition, the high density of oxygen vacancy (i.e., Mo^{5+}) in TMO, which results in low R_{sheet} in MO TFs, degrades their long-term stability through a reaction with airborne molecules such as water when exposed to the ambient environment, thereby lowering the PCE of the fabricated HSC over time [25]. Figure 5a depicts the change in τ_{eff} over time for Si samples deposited with VO or MO for both sides, respectively, and as shown in the graph, for MO, τ_{eff} is measured at 39% of the initial level over the 2000 h measurement. In contrast, in the case of VO with a relatively low oxygen vacancy density, τ_{eff} is maintained at a higher level, which is 81% of the initial level, and comparison of these two results clearly indicate poor stability of MO. The degradation of the TMO over time is more pronounced from the change in HSC efficiency over time, as shown in Figure 5b. After 2000 h, PCE of the MO/Si HSC was measured at 27% of the initial level, whereas that of VO/Si HSC was at 72%. This suggests that excessive formation of high oxygen vacancies in MO considerably degrades the passivation performance and PCE. Thus, for fabrication of high-efficiency TMO/Si HSCs with long-term stability, techniques to simultaneously achieve (1) low TMO resistance; (2) improved Si surface passivation; and (3) low oxygen vacancy density must be developed.

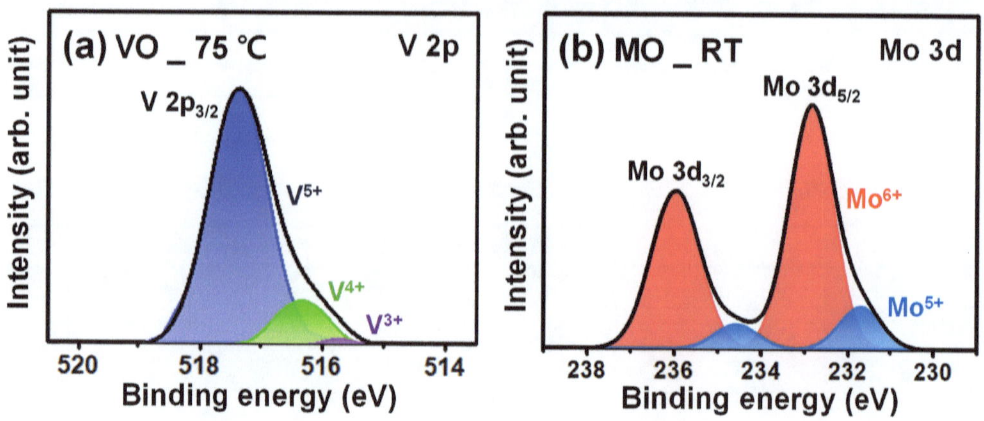

Figure 4. (**a**) V(2p) XPS spectra of VO TF deposited at a substrate temperature of 75 °C and (**b**) Mo(3d) XPS spectra of MO TF deposited at RT.

Table 2. Integrated peak area ratio of vanadium and molybdenum oxidation states of each TMO TF.

Sample	Ratio of V Oxidation State (%)			Sample	Ratio of Mo Oxidation State (%)	
	V^{5+}	V^{4+}	V^{3+}		Mo^{6+}	Mo^{5+}
VO (@ 75 °C)	88.3	10.5	1.2	MO (@ RT)	80.5	19.5

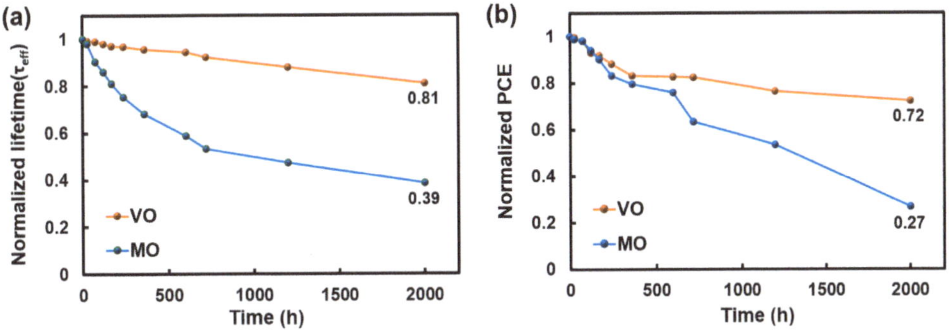

Figure 5. (**a**) Degradation of τ_{eff} and (**b**) PCE degradation of TMO/Si HSCs over 2000 h.

3.2. Effect of TMO Thickness on TMT/Si HSC Performance

As discussed in Section 3.1, the VO TF offers excellent Si surface passivation characteristics, but a high resistance value; in contrast, the MO TF provides a low resistance, but exhibits poor Si surface passivation performance. In this study, we endeavored to build an asymmetric TMT-structured HSC that uses both TMOs (MO and VO) to increase device performance and long-term stability by exploiting the properties of the two TMOs. In the design of the HSC structure, we aimed to further improve the charge carrier extraction efficiency of the fabricated HSC by sandwiching the Ag metal layer as a conductive layer between the two TMO TFs [28,29]. For the fabrication of the asymmetric TMT/Si HSC, a low R_{sheet} MO TF was applied as the BTMO, and a VO TF with a low oxygen vacancy density was deposited as the TTMO to overcome the insufficient Si surface passivation of the bottom MO TF through the enhanced field-effect passivation. The thickness of each TMO TF was optimized to fabricate an asymmetric TMT/Si HSC; to this end, MO/Si HSCs were first fabricated, and the effect was analyzed according to the thickness of the bottom MO TF. Figure 6 depicts J–V curves of the MO/Si HSCs with 8-, 15-, and 30-nm-thick MO TFs under dark and illuminated conditions. The solar cell parameters of the fabricated HSCs are outlined in Table 3. The highest PCE, 3.81%, was achieved when the 8-nm thick MO TF was used, and the PCE declined with increasing thickness of the MO TF. The drop in PCE is primarily attributed to increased R_s with 15- and 30-nm-thick MO TFs, 3.39 and 3.57, respectively, as shown in Table 3.

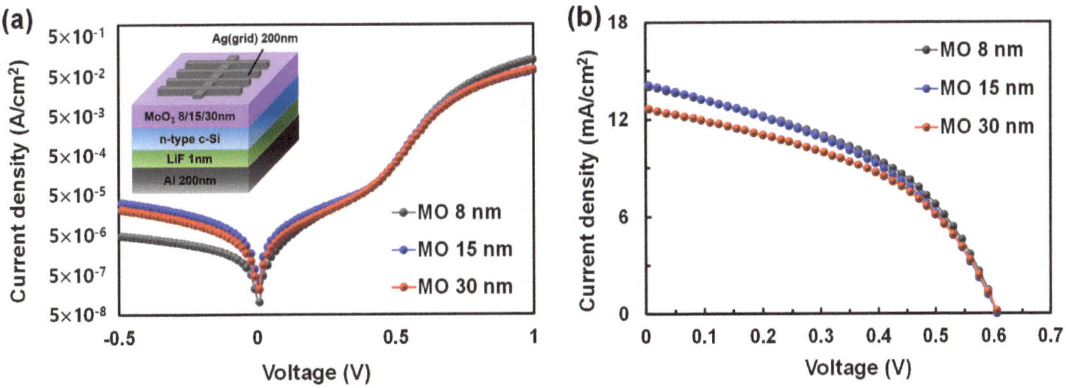

Figure 6. J–V curves of the MO/Si HSCs according to MO layer thickness under (**a**) dark and (**b**) light (AM1.5) conditions.

Table 3. Solar cell characteristic parameters of the MO/Si HSCs according to the MO layer thickness under light (AM1.5) conditions.

MO Thickness	J_{sc} (mA/cm^2)	V_{oc} (mV)	FF (%)	R_{sh} (Ω·cm^2)	R_s (Ω·cm^2)	PCE (%)
8 nm	14.07	606	44.73	14,089	2.10	3.81 ± 0.05
15 nm	14.16	605	42.84	14,166	3.39	3.67 ± 0.06
30 nm	12.68	607	45.16	12,666	3.57	3.48 ± 0.09

Furthermore, the effect of the top VO thickness on TMT/Si HSC performance was investigated. To this end, the TMT/Si HSCs were fabricated with the bottom MO and middle Ag layer thicknesses fixed at 8 nm and 15 nm, respectively, and the top VO with varying thicknesses of 15 nm, 35 nm, 55 nm, and 75 nm was deposited. The changes in solar cell parameters and τ_{eff} according to the top layer thickness were analyzed. Figure 7 and Table 4 show the J–V curves and solar cell parameters under dark and illuminated conditions as a function of the thickness of the top VO, respectively. The values in Table 4 indicate that as the thickness of the TTMO increased, the PCE of the fabricated device increased. The highest PCE (7.57%) was achieved when the top VO has a thickness of 55 nm. However, the sample with 75 nm of top VO TF showed a lower PCE (3.33%). This drop in PCE is mainly attributed to the increased R_s, 12.46 Ωcm^2, with 75 nm compared to 2.91 Ωcm^2 of 55 nm as shown in Table 4.

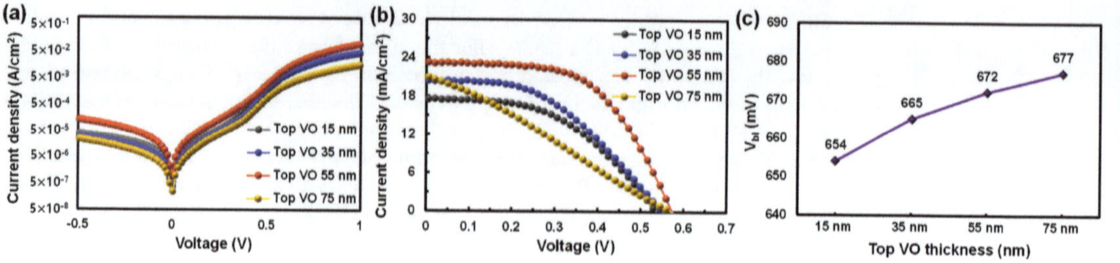

Figure 7. J–V curves of the TMT/Si HSCs according to the top VO layer thickness under (**a**) dark and (**b**) light (AM1.5) conditions. (**c**) V_{bi} values of the fabricated TMT/Si HSCs according to the top VO layer thickness.

Table 4. Solar cell characteristic parameters of the TMT/Si HSCs according to the top VO layer thickness under light (AM1.5) conditions.

MO/Ag/VO	J_{sc} (mA/cm^2)	V_{oc} (mV)	FF (%)	R_{sh} (Ω·cm^2)	R_s (Ω·cm^2)	PCE (%)
8 nm/15 nm/15 nm	17.54	537	48.36	17,566	6.31	4.56 ± 0.17
8 nm/15 nm/35 nm	20.40	552	45.37	20,394	4.19	5.11 ± 0.14
8 nm/15 nm/55 nm	23.24	574	56.78	23,207	2.91	7.57 ± 0.07
8 nm/15 nm/75 nm	21.08	573	27.57	21,051	12.46	3.33 ± 0.06

The improvement in PCE with the thicker top VO TFs is because the work function of the top VO becomes more dominant at the TMO/Si interface as the thickness increases [30,31]. However, in this TMT structure, the work function of top VO would be inevitably pinned at that of inserted metal (i.e., Ag) layer. Therefore, to expect thickness effect of top VO TF, the deposited Ag layer should not fully cover the MO surface to offer direct contact between top VO and bottom MO TFs over a certain fraction of the area. Therefore, before further discussion on the thickness effect of top VO TF, the morphology of the deposited Ag layer was measured with AFM, and the results are shown in Figure S1. For AFM, three different Ag thicknesses (i.e., 10, 15, 20 nm) were fabricated to investigate

Ag thin film growth mechanisms in addition to their morphologies. From the AFM results, the deposited Ag layer on the MO surface produced discontinuous island-shaped surface morphologies for all three samples. which indicate that dominant thin-film growth mechanism for Ag layer on the MO surface is the Volmer–Weber growth mechanism. Based on these AFM results, it is clear that the top VO TF would form direct contact with the bottom MO TF over a certain area fraction to weaken the fermi-level pinning with Ag. (Note: in Figure S2 and Table S1, the measured J–V curve and the device parameters of TMT/Si HSCs are also provided, which showed that a 15 nm Ag layer is the optimal thickness for HSC fabrication.)

The work function values for MO and VO are known to be similar at 6.9 and 7.0 eV, respectively [16,32]. However, as shown from the XPS results presented in Table 2, the stochiometric V_2O_5 ratio of VO is higher than the ratio of the oxidation state of MO; thus, its work function value is also predicted to be higher than that of MO. This is because the reduction of the oxygen vacancy density induces a decrease in the free carrier concentration inside the TMO, thereby weakening the n-type doping effect [33,34]. Therefore, with the increasing thickness of the top VO, VO work function value increases; a higher value is predicted to dominate the behavior at the TMO/Si interface. Consequently, as the thickness of the top VO increases, the built-in potential (V_{bi}) of the TMO/Si interface increases, and this increase may lead to (1) reduced surface recombination loss because of the strengthened field-effect passivation effect; and (2) higher open circuit voltage (V_{oc}) for the fabricated HSCs [35,36]. To validate the assumption, the current density–voltage diode equation (Equation (1)) was applied to the J–V curve (Figure 7a) obtained from the fabricated HSCs under the dark condition, and the saturation current (J_0) was calculated from Equation (2):

$$J = J_0\left(\exp\left(\frac{eV}{nkT}\right) - 1\right); \tag{1}$$

$$J_0 = A^* AT^2 \exp\left(-\frac{V_{bi}}{kT}\right), \tag{2}$$

where A is the contact area; A^* is the effective Richardson constant (120 $Acm^{-2}K^{-2}$ for n-type silicon); T is 25 °C (298 K); k is the Boltzmann constant; n is the ideality factor; J_0 is the reverse saturation current density; V_{bi} is the barrier height in Schottky diodes (i.e., a built-in-potential); and q is the elementary charge. In this way, V_{bi} was extracted and presented in Figure 7c [37]. In Figure 7c, it can be seen that the calculated V_{bi} increased with the increasing top VO thickness. The increase in the V_{oc} observed in Figure 7c reflects and confirms the influence of V_{bi} on the solar cell parameters. Furthermore, the higher V_{bi} induces an imbalance in electron and hole concentrations owing to the strengthened field-effect at the TMO/Si interface—this is expected to reduce the carrier loss caused by the Shockley–Read–Hall recombination (R_{SRH}) at the Si surface as expressed in Equation (3); this, in turn, extends the τ_{eff} of the device [37,38].

$$R_{SRH} = \left(n_s p_s - n_i^2\right) v_{th} \times \int_{E_v}^{E_c} \frac{D(E_t)}{(n_s + n_1)/\sigma_p(E_t) + (p_s + p_1)/\sigma_n(E_t)} dE_t, \tag{3}$$

where n_1 and p_1 are the densities of the electrons and holes in the bulk, respectively; n_s and p_s are the densities of the electrons and holes at the surface, respectively; vth is the thermal velocity; E_c and E_v are the conduction and valance band energies, respectively; D_{it} is the density of the interface states; and σ_n and σ_p are the energy-dependent capture cross sections of the holes and electrons, respectively. Figure 8 shows the measured τ_{eff} values as a function of the top VO thickness. The τ_{eff} value increased as the top VO thickness increased, as predicted. The τ_{eff} was recorded its highest value (155.2 µs) at a thickness of 55 nm, and it decreased somewhat at 75 nm but still remained high. Thus, for the fabrication of the TMT/Si HSC with different TMOs, increasing the TTMO thickness up to a value below the critical thickness (the thickness at which the device resistance increases) is expected to improve the overall PCE of the HSC.

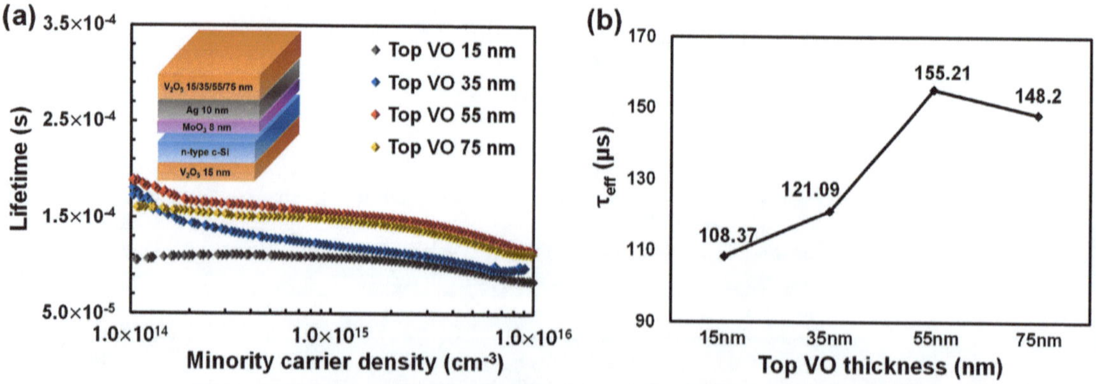

Figure 8. (**a**) Minority carrier lifetimes of the TMT multilayer deposited with various top VO thicknesses; (**b**) τ_{eff} values at the carrier injection level of 1.5×10^{15} cm^{-3}.

3.3. Comparison between Asymmetric and Symmetric TMT/Si HSCs

In this section, we compared the solar cell performance and long-term stability of the asymmetric TMT/Si HSC with the conventional symmetric TMT/Si HSC having one kind of TMOs for BTMO and TTMO. For the symmetric TMT/Si HSCs, samples for comparison were fabricated with structures of MO (8 nm)/Ag (15 nm)/MO (55 nm) (MAM) and VO (8 nm)/Ag (15 nm)/VO (55 nm) (VAV). For the asymmetric TMT/Si HSCs, samples were fabricated with structures of MO (8 nm)/Ag (15 nm)/VO (55 nm) (MAV) and VO (8 nm)/Ag (15 nm)/MO (55 nm) (VAM). Figure 9 shows the J–V curves of the fabricated HSCs under dark and illuminated conditions, and the solar cell parameters of each type of HSCs are presented in Table 5.

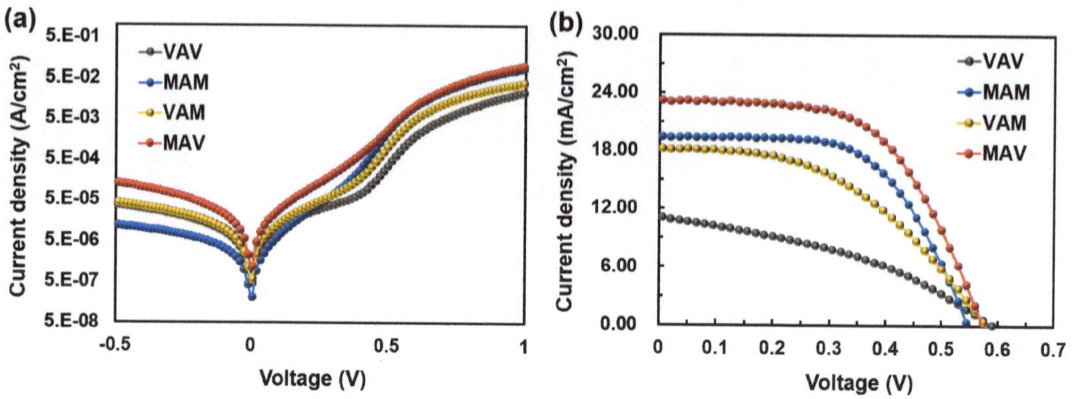

Figure 9. J–V curves of the fabricated TMT/Si HSCs under (**a**) dark and (**b**) light (AM1.5) conditions.

Table 5. Solar cell characteristic parameters of the fabricated TMT/Si HSCs under light (AM 1.5) conditions.

Sample Structures	J_{sc} (mA/cm^2)	V_{oc} (mV)	FF (%)	R_{sh} ($\Omega\cdot cm^2$)	R_s ($\Omega\cdot cm^2$)	PCE (%)
VO 8 nm/Ag 15 nm/VO 55 nm (VAV)	11.11	591	37.76	11,146	5.47	2.48 ± 0.05
MO 8 nm/Ag 15 nm/MO 55 nm (MAM)	19.42	547	59.22	19,411	2.54	6.29 ± 0.22
VO 8 nm/Ag 15 nm/MO 55 nm (VAM)	18.20	579	45.90	18,313	7.85	4.84 ± 0.18
MO 8 nm/Ag 15 nm/VO 55 nm (MAV)	23.24	574	56.78	23,207	2.91	7.57 ± 0.07

Among the four different TMT/Si HSCs, the HSCs that applied MO as the BTMO, MAM and MAV, exhibited higher PCEs compared to those of VAV and VAM. This is attributed to an increase in the charge extraction efficiency of the photogenerated carrier, resulting from the low resistance of the MO layer itself. However, in the case of MAM (where MO was also applied as the TTMO), as shown in Figure 10, due to the formation of excessive oxygen vacancies in the MO layer, rapid PCE degradation over time was observed in relation to other HSCs used for comparison; this reveals the undesirable poor long-term stability of the HSC. However, in the case of the asymmetric TMT/Si HSC with the MAV structure, with the application of MO as the BTMO and VO as the TTMO, the highest PCE of 7.57% was achieved; enhanced long-term stability was also observed. From these results, in the future, a high-efficiency TMT/Si HSC with long-term stability can be developed through additional stoichiometry control and thickness optimization of the TMOs.

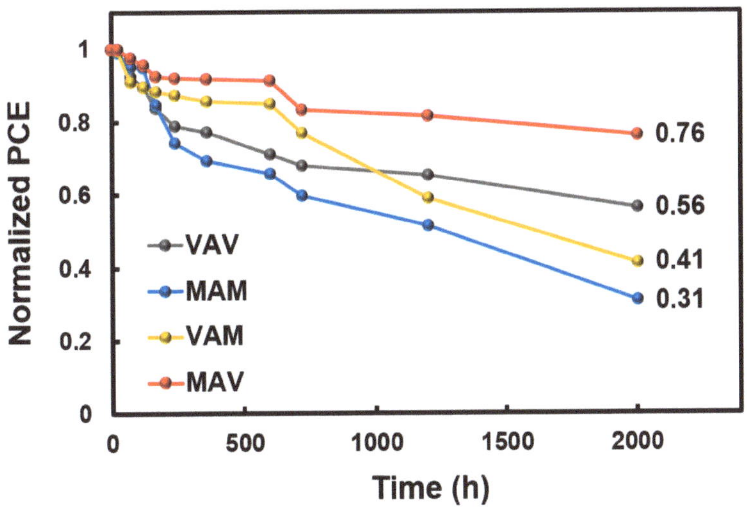

Figure 10. PCE degradation ratios of the fabricated TMT/Si HSCs over 2000 h.

4. Conclusions

In this study, an asymmetric TMT (TMO–Metal–TMO) structure was introduced to overcome the limitations of existing TMO/Si HSCs. The asymmetric TMT structure was achieved by depositing MO with low resistance as the BTMO and VO with improved field-effect passivation as the TTMO. The results showed that the oxygen vacancy density plays a crucial role in determining the electrical properties of TMO TFs, the Si surface passivation effect, and the long-term stability of the HSC. The VO TF demonstrated better Si surface passivation and long-term stability, but its high resistance resulted in a low PCE compared to the MO TF. By combining the advantages of both MO and VO, the asymmetric TMT/Si HSC showed 20% improved PCE and long-term stability compared to the conventional symmetric TMT/Si HSC. This is because (1) lower resistance with bottom MO having higher density of oxygen vacancy; and (2) the enhanced field effect passivation and long-term stability with top VO TF, revealing lower density of oxygen vacancy. Therefore, the proposed asymmetric TMT/Si HSC could be a promising solution for low-cost and high-efficiency c-Si-based solar cells with improved Si surface passivation and long-term stability.

Supplementary Materials: The following supporting information can be downloaded at: https://www.mdpi.com/article/10.3390/ma16165550/s1, Figure S1: AFM images of Ag layers deposited on MO surface: (a) 10 nm, (b) 15 nm, and (c) 20 nm; Figure S2: J–V curves of the TMT/Si HSCs according to the middle Ag layer thickness under (a) dark and (b) illumination (AM1.5) conditions; Table S1: Solar cell characteristic parameters of the TMT/Si HSCs according to the middle Ag layer thickness under illumination (AM1.5) conditions.

Author Contributions: Conceptualization, J.-Y.C.; Methodology, Y.-J.Y., Y.-K.K., J.H.L. and J.H.S.; Validation, J.-Y.C.; Formal analysis, J.-Y.C.; Investigation, Y.-C.J. and J.-Y.C.; Resources, J.-Y.C.; Data curation, Y.-C.J., Y.-J.Y., J.H.S. and J.-Y.C.; Writing—original draft, Y.-C.J.; Writing—review & editing, J.-Y.C.; Visualization, Y.-C.J.; Project administration, J.-Y.C.; Funding acquisition, J.-Y.C. All authors have read and agreed to the published version of the manuscript.

Funding: This work was supported by the National Research Foundation of Korea (NRF) Grant 2020R1F1A1053556 funded by the Korean Ministry of Science and ICT.

Institutional Review Board Statement: Not applicable.

Informed Consent Statement: Not applicable.

Data Availability Statement: Not applicable.

Conflicts of Interest: The authors declare no conflict of interest.

References

1. Yoshikawa, K.; Kawasaki, H.; Yoshida, W.; Irie, T.; Konishi, K.; Nakano, K.; Uto, T.; Adachi, D.; Kanematsu, M.; Uzu, H. Silicon heterojunction solar cell with interdigitated back contacts for a photoconversion efficiency over 26%. *Nat. Energy* **2017**, *2*, 17032. [CrossRef]
2. Wilson, G.M.; Al-Jassim, M.; Metzger, W.K.; Glunz, S.W.; Verlinden, P.; Xiong, G.; Mansfield, L.M.; Stanbery, B.J.; Zhu, K.; Yan, Y. The 2020 photovoltaic technologies roadmap. *J. Phys. D Appl. Phys.* **2020**, *53*, 493001. [CrossRef]
3. Dai, H.; Yang, L.; He, S. <50-μm thin crystalline silicon heterojunction solar cells with dopant-free carrier-selective contacts. *Nano Energy* **2019**, *64*, 103930.
4. Gao, P.; Ding, K.; Wang, Y.; Ruan, K.; Diao, S.; Zhang, Q.; Sun, B.; Jie, J. Crystalline Si/graphene quantum dots heterojunction solar cells. *J. Phys. Chem. C* **2014**, *118*, 5164–5171. [CrossRef]
5. Cruz, A.; Wang, E.-C.; Morales-Vilches, A.B.; Meza, D.; Neubert, S.; Szyszka, B.; Schlatmann, R.; Stannowski, B. Effect of front TCO on the performance of rear-junction silicon heterojunction solar cells: Insights from simulations and experiments. *Sol. Energy Mater. Sol. Cells* **2019**, *195*, 339–345. [CrossRef]
6. Niemelä, J.-P.; Macco, B.; Barraud, L.; Descoeudres, A.; Badel, N.; Despeisse, M.; Christmann, G.; Nicolay, S.; Ballif, C.; Kessels, W.M. Rear-emitter silicon heterojunction solar cells with atomic layer deposited ZnO: Al serving as an alternative transparent conducting oxide to In_2O_3: Sn. *Sol. Energy Mater. Sol. Cells* **2019**, *200*, 109953. [CrossRef]
7. Gerling, L.G.; Mahato, S.; Morales-Vilches, A.; Masmitja, G.; Ortega, P.; Voz, C.; Alcubilla, R.; Puigdollers, J. Transition metal oxides as hole-selective contacts in silicon heterojunctions solar cells. *Sol. Energy Mater. Sol. Cells* **2016**, *145*, 109–115. [CrossRef]
8. Messmer, C.; Bivour, M.; Schön, J.; Hermle, M. Requirements for efficient hole extraction in transition metal oxide-based silicon heterojunction solar cells. *J. Appl. Phys.* **2018**, *124*, 085702. [CrossRef]
9. He, J.; Wan, Y.; Gao, P.; Tang, J.; Ye, J. Over 16.7% efficiency organic-silicon heterojunction solar cells with solution-processed dopant-free contacts for both polarities. *Adv. Funct. Mater.* **2018**, *28*, 1802192. [CrossRef]

10. Rafique, S.; Abdullah, S.M.; Shahid, M.M.; Ansari, M.O.; Sulaiman, K. Significantly improved photovoltaic performance in polymer bulk heterojunction solar cells with graphene oxide/PEDOT: PSS double decked hole transport layer. *Sci. Rep.* **2017**, *7*, 39555. [CrossRef]
11. Liu, Y.; Chen, Q.; Duan, H.-S.; Zhou, H.; Yang, Y.M.; Chen, H.; Luo, S.; Song, T.-B.; Dou, L.; Hong, Z. A dopant-free organic hole transport material for efficient planar heterojunction perovskite solar cells. *J. Mater. Chem. A* **2015**, *3*, 11940–11947. [CrossRef]
12. Liu, Y.; Hong, Z.; Chen, Q.; Chen, H.; Chang, W.H.; Yang, Y.; Song, T.B. Perovskite solar cells employing dopant-free organic hole transport materials with tunable energy levels. *Adv. Mater.* **2016**, *28*, 440–446. [CrossRef] [PubMed]
13. Wu, W.; Bao, J.; Jia, X.; Liu, Z.; Cai, L.; Liu, B.; Song, J.; Shen, H. Dopant-free back contact silicon heterojunction solar cells employing transition metal oxide emitters. *Phys. Status Solidi (RRL) Rapid Res. Lett.* **2016**, *10*, 662–667. [CrossRef]
14. Mehmood, H.; Nasser, H.; Tauqeer, T.; Turan, R. Simulation of silicon heterostructure solar cell featuring dopant-free carrier-selective molybdenum oxide and titanium oxide contacts. *Renew. Energy* **2019**, *143*, 359–367. [CrossRef]
15. Almora, O.; Gerling, L.G.; Voz, C.; Alcubilla, R.; Puigdollers, J.; Garcia-Belmonte, G. Superior performance of V_2O_5 as hole selective contact over other transition metal oxides in silicon heterojunction solar cells. *Sol. Energy Mater. Sol. Cells* **2017**, *168*, 221–226. [CrossRef]
16. Gerling, L.G.; Mahato, S.; Voz, C.; Alcubilla, R.; Puigdollers, J. Characterization of transition metal oxide/silicon heterojunctions for solar cell applications. *Appl. Sci.* **2015**, *5*, 695–705. [CrossRef]
17. Masmitjà, G.; Ortega, P.; Puigdollers, J.; Gerling, L.; Martín, I.; Voz, C.; Alcubilla, R. Interdigitated back-contacted crystalline silicon solar cells with low-temperature dopant-free selective contacts. *J. Mater. Chem. A* **2018**, *6*, 3977–3985. [CrossRef]
18. Lu, M.; Das, U.; Bowden, S.; Hegedus, S.; Birkmire, R. Optimization of interdigitated back contact silicon heterojunction solar cells: Tailoring hetero-interface band structures while maintaining surface passivation. *Prog. Photovolt. Res. Appl.* **2011**, *19*, 326–338. [CrossRef]
19. Yu, C.; Xu, S.; Yao, J.; Han, S. Recent advances in and new perspectives on crystalline silicon solar cells with carrier-selective passivation contacts. *Crystals* **2018**, *8*, 430. [CrossRef]
20. Gerling, L.G.; Masmitja, G.; Ortega, P.; Voz, C.; Alcubilla, R.; Puigdollers, J. Passivating/hole-selective contacts based on V2O5/SiOx stacks deposited at ambient temperature. *Energy Procedia* **2017**, *124*, 584–592. [CrossRef]
21. Gerling, L.G.; Voz, C.; Alcubilla, R.; Puigdollers, J. Origin of passivation in hole-selective transition metal oxides for crystalline silicon heterojunction solar cells. *J. Mater. Res.* **2017**, *32*, 260–268. [CrossRef]
22. Young, D.L.; Nemeth, W.; Grover, S.; Norman, A.; Lee, B.G.; Stradins, P. Carrier-selective, passivated contacts for high efficiency silicon solar cells based on transparent conducting oxides. In Proceedings of the 2014 IEEE 40th Photovoltaic Specialist Conference (PVSC), Denver, CO, USA, 8–13 June 2014; pp. 1–5.
23. Feldmann, F.; Simon, M.; Bivour, M.; Reichel, C.; Hermle, M.; Glunz, S. Carrier-selective contacts for Si solar cells. *Appl. Phys. Lett.* **2014**, *104*, 181105. [CrossRef]
24. Jeong, G.S.; Jung, Y.-C.; Park, N.Y.; Yu, Y.-J.; Lee, J.H.; Seo, J.H.; Choi, J.-Y. Stoichiometry and Morphology Analysis of Thermally Deposited V_2O_{5-x} Thin Films for Si/V_2O_{5-x} Heterojunction Solar Cell Applications. *Materials* **2022**, *15*, 5243. [CrossRef]
25. Gao, P.; Yang, Z.; He, J.; Yu, J.; Liu, P.; Zhu, J.; Ge, Z.; Ye, J. Dopant-free and carrier-selective heterocontacts for silicon solar cells: Recent advances and perspectives. *Adv. Sci.* **2018**, *5*, 1700547. [CrossRef]
26. Dauwe, S.; Mittelstädt, L.; Metz, A.; Hezel, R. Experimental evidence of parasitic shunting in silicon nitride rear surface passivated solar cells. *Prog. Photovolt. Res. Appl.* **2002**, *10*, 271–278. [CrossRef]
27. Reeves, G.; Harrison, H. Obtaining the specific contact resistance from transmission line model measurements. *IEEE Electron Device Lett.* **1982**, *3*, 111–113. [CrossRef]
28. Ros Costals, E. Transparent electrodes based on ultra-thin dielectric-metal-dielectric multilayer structures. Master's Thesis, Universitat Politècnica de Catalunya, Barcelona, Spain, 2019.
29. Akdemir, O.; Zolfaghari Borra, M.; Nasser, H.; Turan, R.; Bek, A. MoOx/Ag/MoOx multilayers as hole transport transparent conductive electrodes for n-type crystalline silicon solar cells. *Int. J. Energy Res.* **2020**, *44*, 3098–3109. [CrossRef]
30. Hornauer, H.; Vancea, J.; Reiss, G.; Hoffmann, H. Thickness dependence of the work function in double-layer metallic films. *Z. Für Phys. B Condens. Matter* **1989**, *77*, 399–407. [CrossRef]
31. Kim, J.H.; Lee, J.; Kim, J.H.; Hwang, C.; Lee, C.; Park, J.Y. Work function variation of MoS2 atomic layers grown with chemical vapor deposition: The effects of thickness and the adsorption of water/oxygen molecules. *Appl. Phys. Lett.* **2015**, *106*, 251606. [CrossRef]
32. Meyer, J.; Hamwi, S.; Kröger, M.; Kowalsky, W.; Riedl, T.; Kahn, A. Transition metal oxides for organic electronics: Energetics, device physics and applications. *Adv. Mater.* **2012**, *24*, 5408–5427. [CrossRef]
33. Li, Y.; Phattalung, S.N.; Limpijumnong, S.; Kim, J.; Yu, J. Formation of oxygen vacancies and charge carriers induced in the n-type interface of a $LaAlO_3$ overlayer on $SrTiO_3$ (001). *Phys. Rev. B* **2011**, *84*, 245307. [CrossRef]
34. Liu, L.; Mei, Z.; Tang, A.; Azarov, A.; Kuznetsov, A.; Xue, Q.-K.; Du, X. Oxygen vacancies: The origin of n-type conductivity in ZnO. *Phys. Rev. B* **2016**, *93*, 235305. [CrossRef]
35. Glunz, S.W.; Biro, D.; Rein, S.; Warta, W. Field-effect passivation of the SiO_2Si interface. *J. Appl. Phys.* **1999**, *86*, 683–691. [CrossRef]
36. He, J.; Gao, P.; Ling, Z.; Ding, L.; Yang, Z.; Ye, J.; Cui, Y. High-efficiency silicon/organic heterojunction solar cells with improved junction quality and interface passivation. *ACS Nano* **2016**, *10*, 11525–11531. [CrossRef]

37. Park, N.Y.; Jeong, G.S.; Yu, Y.-J.; Jung, Y.-C.; Lee, J.H.; Seo, J.H.; Choi, J.-Y. Photovoltaic Device Application of a Hydroquinone-Modified Conductive Polymer and Dual-Functional Molecular Si Surface Passivation Technology. *Polymers* **2022**, *14*, 478. [CrossRef]
38. McIntosh, K.R.; Black, L.E. On effective surface recombination parameters. *J. Appl. Phys.* **2014**, *116*, 014503. [CrossRef]

materials

MDPI

Article

Tailoring the Structure and Properties of Epitaxial Europium Tellurides on Si(100) through Substrate Temperature Control

Fan Yu [1], Xiaodong Qiu [1], Jinming Zhou [2], Lin Huang [1], Bin Yang [1], Junming Liu [1,3], Di Wu [1,3], Gan Wang [2,4] and Yi Zhang [1,3,5,*]

1 National Laboratory of Solid State Microstructure, School of Physics, Nanjing University, Nanjing 210093, China; dg1822056@smail.nju.edu.cn (F.Y.)
2 Department of Physics, and Shenzhen Institute for Quantum Science and Engineering, Southern University of Science and Technology, Shenzhen 518055, China
3 Collaborative Innovation Center of Advanced Microstructures, Nanjing University, Nanjing 210093, China
4 Guangdong Provincial Key Laboratory of Quantum Science and Engineering, Shenzhen 518055, China
5 Hefei National Laboratory, Hefei 230088, China
* Correspondence: zhangyi@nju.edu.cn

Abstract: In this study, we improved the growth procedure of EuTe and realized the epitaxial growth of $EuTe_4$. Our research demonstrated a selective growth of both EuTe and $EuTe_4$ on Si(100) substrates using the molecular beam epitaxy (MBE) technique and reveals that the substrate temperature plays a crucial role in determining the structural phase of the grown films: EuTe can be obtained at a substrate temperature of 220 °C while lowering down the temperature to 205 °C leads to the formation of $EuTe_4$. A comparative analysis of the transmittance spectra of these two films manifested that EuTe is a semiconductor, whereas $EuTe_4$ exhibits charge density wave (CDW) behavior at room temperature. The magnetic measurements displayed the antiferromagnetic nature in EuTe and $EuTe_4$, with Néel temperatures of 10.5 and 7.1 K, respectively. Our findings highlight the potential for controllable growth of EuTe and $EuTe_4$ thin films, providing a platform for further exploration of magnetism and CDW phenomena in rare earth tellurides.

Keywords: EuTe; $EuTe_4$; molecular beam epitaxy; thin film growth; structural characterization

Citation: Yu, F.; Qiu, X.; Zhou, J.; Huang, L.; Yang, B.; Liu, J.; Wu, D.; Wang, G.; Zhang, Y. Tailoring the Structure and Properties of Epitaxial Europium Tellurides on Si(100) through Substrate Temperature Control. *Materials* **2023**, *16*, 7093. https://doi.org/10.3390/ma16227093

Academic Editors: Sebastiano Vasi and Ulderico Wanderlingh

Received: 30 September 2023
Revised: 31 October 2023
Accepted: 6 November 2023
Published: 9 November 2023

1. Introduction

Rare earth tellurides ($ReTe_x$) exhibit a diverse range of intriguing properties including charge density waves (CDW) [1–4], two-dimensional (2D) magnetism [5,6], as well as thermal hysteresis effects on resistivity and CDW gap [7,8], making them promising candidates for studying electron correlation phenomena and various applications in spintronics. Among these materials, europium tellurides are particularly unique due to the half-filled $4f$ orbital of the Eu atom ([Xe] $4f^7$ $6s^2$). According to Hund's rule, the Eu atom possesses the maximum spin angular momentum with zero orbital angular momentum ($L = 0, J = S = \frac{7}{2}$). As a result, the magnetism in europium tellurides arises solely from the spin of the Eu atom.

Europium chalcogenides were among the earliest discovered magnetic semiconductors. In particular, EuTe, a renowned antiferromagnetic semiconductor, has been extensively studied for over half a century [9–12]. As shown in Figure 1a, it exhibits a face-centered cubic rock salt structure with a bulk lattice constant of $a = b = c = 6.598$ Å [13]. The valence state of Eu and Te in EuTe are +2 and −2, respectively [14]. In EuTe, the magnetic moments are carried by Eu, and the magnetic properties of EuTe can be depicted by an isotropic Hamiltonian, accounting only for the nearest and next-nearest neighbor exchange interactions [15–17]. At low temperatures, EuTe transitions into a type II antiferromagnet with a Néel temperature of 9.8 K [18].

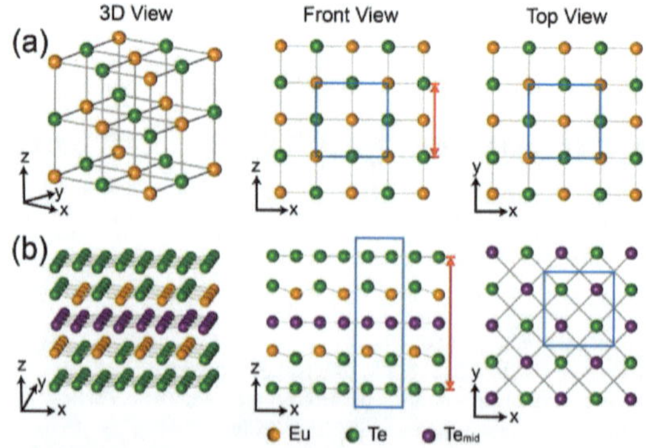

Figure 1. Three-dimensional (**left** panels), front (**middle** panels), and top views (**right** panels) of the (**a**) EuTe and (**b**) $EuTe_4$ lattice. The yellow balls represent the Eu atoms, the purple balls represent the Te atoms in the middle layer (Te_{mid}) of $EuTe_4$, and the green balls represent the Te atoms in other positions. The blue solid rectangles indicate the respective unit cells, while the red double-headed arrows denote the single-layer thickness of both films.

In contrast, $EuTe_4$ is a newly discovered van der Waals layered material that has received significant attraction in recent years [7,8,19–21]. As illustrated in Figure 1b, $EuTe_4$ adopts an orthorhombic lattice structure with a space group of Pmmn (No. 59) at room temperature. The lattice constant of $EuTe_4$ in its normal state are as follows: a = 4.5119(2) Å, b = 4.6347(2) Å, c = 15.6747(10) Å [20]. The unit cell of $EuTe_4$ comprises a Te-EuTe-Te-EuTe-Te quintuple layer. The valence states of the Eu and Te ions in the Eu-Te slab are +2 and −2, respectively, while the valence state of the isolated Te layers remains nominally neutral [7]. The nearly square Te layers are unstable and tend to be distorted, resulting in CDW transition above 400 K [7]. Remarkably, $EuTe_4$ exhibits a unique type of metastability, characterized by a thermal hysteresis that spans over 400 K in temperature. More specifically, the CDW gap and electrical resistivity of $EuTe_4$ manifest different behaviors even at an identical temperature, depending upon the preceding temperature variation path (for instance, whether it was heated up to 300 K or cooled down to 300 K) [7]. The origin of this thermal hysteresis deviates from conventional mechanisms and can be elucidated by the switching of CDW phases in distinct Te layers, a phenomenon not present in 2D or strongly correlated 3D systems [7].

Despite extensive research on EuTe, the high-quality synthesis of EuTe thin films remains a challenge. Previous studies used BaF_2(111) as the growth substrate, but the large lattice mismatch between EuTe film and BaF_2 substrate necessitated the incorporation of PbTe(111) film as a buffer layer [22–26]. Moreover, the film quality was highly sensitive to the substrate temperature and required a rigorous flux ratio control. Furthermore, the thickness of EuTe film grown on BaF_2 was also limited to 45 layers due to the formation of a strain-induced three-dimensional island [22]. Therefore, it is necessary to improve the growth procedure of EuTe in order to attain a deeper comprehension of the rich magnetic properties within this system. On the other hand, the current research on $EuTe_4$ primarily focuses on its bulk properties, lacking a systematical investigation on $EuTe_4$ thin film in a 2D limit. The synthesis of 2D epitaxial $EuTe_4$ thin film serves as a platform to facilitate our understanding of the mechanisms behind its CDW behavior in subsequent studies. Also, it offers an opportunity to delve into the competitive interactions between different Te atomic layers, unraveling the underlying mechanisms that drive the thermal hysteresis phenomena.

Substrate temperature is a key factor in the molecular beam epitaxial (MBE) growth of thin films. Typically, substrate temperature only affects the morphology and quality of the films [27–29]. In specific instances, such as MoS_2 [30], WSe_2 [31], and $TaTe_2$ [32], precise temperature control enables the selective growth of films with different crystalline structures. By carefully tuning the substrate temperature, one can manipulate the microstructure of the material at the atomic level. Such control paves new paths to precisely tailor the optical, magnetic, and electronic properties of the films.

In this research, we improved the growth procedure for EuTe and realized the epitaxial growth of $EuTe_4$ utilizing MBE. The selective growth of EuTe and $EuTe_4$ was also achieved by precisely adjusting the substrate temperature. In combination with reflection high energy electron diffraction (RHEED), X-ray diffraction (XRD), and scanning transmittance electron microscopy (STEM) techniques, we examined the difference of lattice structures and crystalline orientations between EuTe and $EuTe_4$ films. In addition, we compared the relative stoichiometry ratio and valence state between EuTe and $EuTe_4$ films via X-ray photoelectron spectroscopy (XPS). The experimental data demonstrated that a substrate temperature of 220 °C results in the growth of EuTe, and conversely, $EuTe_4$ film forms at a lower temperature of 205 °C. Additionally, we further investigated the physical properties of the two materials. The XPS spectra near the Fermi level indicated that EuTe is a semiconductor, with its valence band top located about 0.6 eV below the Fermi level. For $EuTe_4$, the density of states stretches to the vicinity of the Fermi level. The transmittance spectra verified the semiconductive property of EuTe and discovered the existence of a CDW gap in $EuTe_4$ at room temperature. In addition, superconducting quantum interference device (SQUID) measurements denoted that EuTe and $EuTe_4$ are both antiferromagnetic materials, with Néel temperatures of 10.5 and 7.1 K, respectively. Our results developed the fabrication and physical property investigation of epitaxial 2D materials based on rare earth elements.

2. Methods

The growth of EuTe and $EuTe_4$ films was conducted in an MBE system (GC inno, Changzhou, Jiangsu, China) with a base pressure of 1×10^{-10} mbar. The conductive Si(100) wafers (n-type boron doped, 0.01~0.05 Ω·cm, HF-Kejing, Hefei, Anhui, China) were selected as substrates. Prior to the growth, the substrates underwent a degassing process at 600 °C for 3 h, followed by a standard flash procedure at 1200 °C to achieve an atomic flat surface [33]. The films were grown by co-deposited high-purity Eu (99.9%) and Te (99.999%) shots (PrMat, Shanghai, China) via standard Knudsen Cells on the Si(100) substrate. The temperatures of the evaporation sources for Eu and Te were maintained at 460 °C and 320 °C, respectively, with flux ratio of Eu:Te keeping ~1:20. The growth of the film was monitored by an in situ RHEED and the growth rate of EuTe and $EuTe_4$ was about 0.3 and 0.2 nm per minute (nm/min), respectively. The thickness of the grown film, defined as the length in the z-direction of Figure 1a,b, was roughly estimated by the growth time.

The crystal structure of the grown films was determined by an ex situ XRD (D8 ADVANCE, Bruker, Billerica, MA, USA) with Cu Kα source (wavelength λ = 1.5406 Å). A spherical aberration-corrected scanning transmission electron microscopy (STEM, Titan Themis G2, FEI, Hillsboro, OR, USA) was employed for further examining the structure of the grown films. To protect the films from possible oxidation in atmosphere and ensure the sample was grounded during the STEM measurements, a ~20 nm thick amorphous Eu metal film was deposited on the sample surface at room temperature before moving the sample from the MBE chamber. The samples were fabricated by the focused ion beam (FIB, Helios Nanolab 600i, FEI, Hillsboro, OR, USA) technique before STEM characterizations. The stoichiometric information of EuTe and $EuTe_4$ were compared by an in situ XPS with a resolution of ~0.2 eV, where the monochromatic X-ray (Al Kα, 1486.7 eV) was used as the excitation light source (Scienta Omicron MECS, Taunusstein, Hesse, Germany). The ex situ transmittance spectra of EuTe and $EuTe_4$ were measured at room temperature, with light incident perpendicular to the sample surface and data collected by a Fourier transform spectrometer (Vertex 80 V, Bruker, Billerica, MA, USA). The magnetic properties were characterized by an ex situ superconducting quantum interference device vibrating sample magnetometer (SQUID-VSM, Quantum Design, San Diego, CA, USA). During the magnetic measurement, a magnetic field of 5000 Oe was applied parallel to the film surface.

3. Results and Discussions

3.1. Growth and Structural Characteristics of EuTe and $EuTe_4$ Films

Figure 2a displays the RHEED pattern of a 2 × 1 surface-reconstructed Si(100) substrate after the standard flash procedure, with the electron beam incident along the Si<100> direction. To clarify the lattice orientations of the substrate and grown films, we present the 45°-rotated RHEED diffraction pattern of Figure 2a in Figure 2b, where the electron beam incident is along the Si<110> direction. Figure 2c provides a schematic diagram of the atomic arrangement on the Si(100) surface. The black arrows indicate the incident directions (0° and 45°) of the RHEED electron beam, while the red and blue double-headed arrows correspond to the space between the diffraction stripes indicated in Figure 2a,b, respectively.

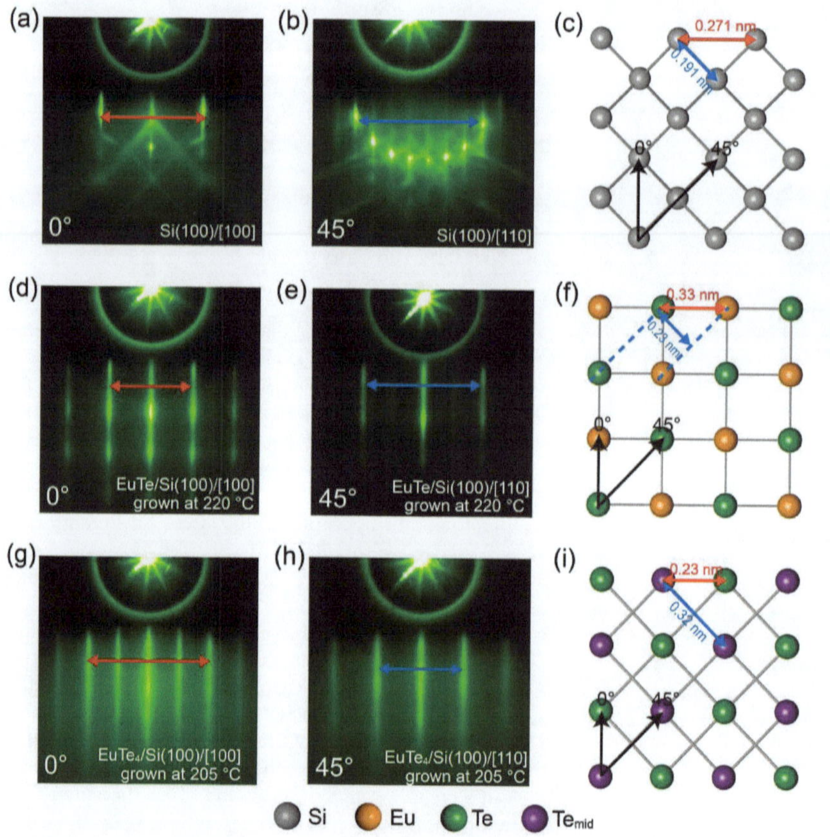

Figure 2. (**a**,**b**) RHEED patterns of a Si(100) substrate with incident beam angles of 0° and 45°, respectively. (**c**) Corresponding top view lattice arrangement for Si(100). The black arrows in (**c**) represent the incident direction of electron beams, while the blue and red arrows between atoms denote the RHEED diffraction stripe spacings as indicated in (**a**,**b**). The in-plane lattice constants, derived from the subsequent STEM analyses, are also annotated in the lattice arrangement diagram of (**c**). (**d**–**i**) Analogous to (**a**–**c**) for (**d**–**f**) a ~10 nm EuTe film and (**g**–**i**) a ~10 nm $EuTe_4$ film.

The substrate temperature played an essential role in determining the structural phase of the grown film. Figure 2d,e present the RHEED diffraction patterns of a ~10 nm thick film grown at 220 °C with the incident electron beam along the Si<100> (0°) and Si<110> (45°) directions, respectively. This film is further identified as EuTe by the subsequent XRD and STEM characterizations. The symmetry of the EuTe RHEED patterns matches that of the silicon substrate, demonstrating a four-fold rotational invariance. This rotational symmetry indicates the grown film possesses a tetragonal structure with the EuTe(001) as the surface orientation, which is notably different from the scenario of EuTe grown on a BaF_2(111) substrate with PbTe as a buffer layer, where the surface orientation of the EuTe/PbTe/BaF_2 film is the (111) plane, displaying a six-fold rotational symmetry [13]. Moreover, as the thickness of the film increases, no evidence of 3D island growth was observed, contrasting with the behavior of the EuTe film on BaF_2(111) substrates, where 3D roughness rapidly increases when reaching the critical layer thickness of 45 layers [24]. This result suggests greater stability and lower binding energy for the EuTe(001) plane [34].

In Figure 2f, the atomic arrangement of the EuTe(001) surface is illustrated, with Eu and Te atoms represented by yellow and green balls, respectively. The RHEED diffraction stripes in Figure 2d (indicated by the red double arrow) correspond to the spacing between adjacent Eu and Te atoms as indicated by the red double arrow in Figure 2f. Similarly, the diffraction stripes in Figure 2e (indicated by the blue double arrow) correspond to the spacing between adjacent Eu(110) and Te(110) planes of EuTe in Figure 2f.

Lowering the substrate temperature to 205 °C results in the growth of $EuTe_4$. Figure 2g,h display the RHEED pattern of a ~10 nm thick $EuTe_4$ film along the Si<100> and <110> directions, respectively. The RHEED diffraction patterns of $EuTe_4$ exhibit a four-fold rotational symmetry, indicating that the thin film's surface orientation is the (001) plane, which is consistent with the cleavage plane of bulk $EuTe_4$ observed in previous angle-resolved photoemission spectroscopic (ARPES) experiments [7,19]. In Figure 2i, a top view of the $EuTe_4$(001) surface is presented, with the green and purple balls representing the topmost and middle Te atomic layers indicated in Figure 1b, respectively. The red and blue double-headed arrows in Figure 2g,h correspond to half of the basis vector along the *x*-axis and the nearest Te atom spacing within a single Te layer, as indicated in Figure 2i, respectively.

A substrate temperature significantly higher or lower than the optimal growth conditions for EuTe and $EuTe_4$ will result in the degeneration of the film quality. More detailed results of the films grown at various temperatures can be seen in Supplementary Material Part A.

We carried out ex situ XRD and STEM characterizations to further identify the crystal structure of the grown films. Figure 3a presents the XRD curves for the Si substrate (black curve), and the films grown at 205 °C (blue curve) and 220 °C (red curve), shown from bottom to top. To display the diffraction peaks with varying intensities on a unified scale, we applied a fourth-root adjustment to the XRD curve intensities. The principal diffraction peaks in each of the three spectra are annotated with their corresponding diffraction indices.

Aside from the dominant peak at 69.40°, which is attributed to the silicon substrate, the XRD diffraction curves of the films grown at 205 °C and 220 °C display notable differences. For the film grown at 220 °C, the peaks at 27.25°, 55.88°, and 89.08° correspond to the EuTe(002), (004), and (006) planes, respectively. This diffraction pattern is distinct from the XRD curve of EuTe(111) grown on the PbTe buffer layer on the BaF_2(111) substrate, where the (222) peak is predominant [13,35]. From the XRD curve, we derived a lattice constant of c = 0.654 nm for EuTe by applying Bragg's law. This value is very close to the lattice constant of 0.650 nm derived from the EuTe film grown on the PbTe layer on the BaF_2(111) substrate [13], confirming that they are the same material but with different crystal orientations.

In contrast to the diffraction pattern of EuTe, $EuTe_4$ exhibits the strongest peak at 28.64°, which is associated with the $EuTe_4$ (005) plane and gives a lattice constant of c = 1.557 nm. This value is consistent with the lattice constant of 1.567 nm obtained from the XRD measurement of the bulk $EuTe_4$ sample [20]. Based on the above XRD curves, we can conclude that the films grown at ~205 °C and ~220 °C belong to different structural phases of $EuTe_4$ and EuTe, respectively.

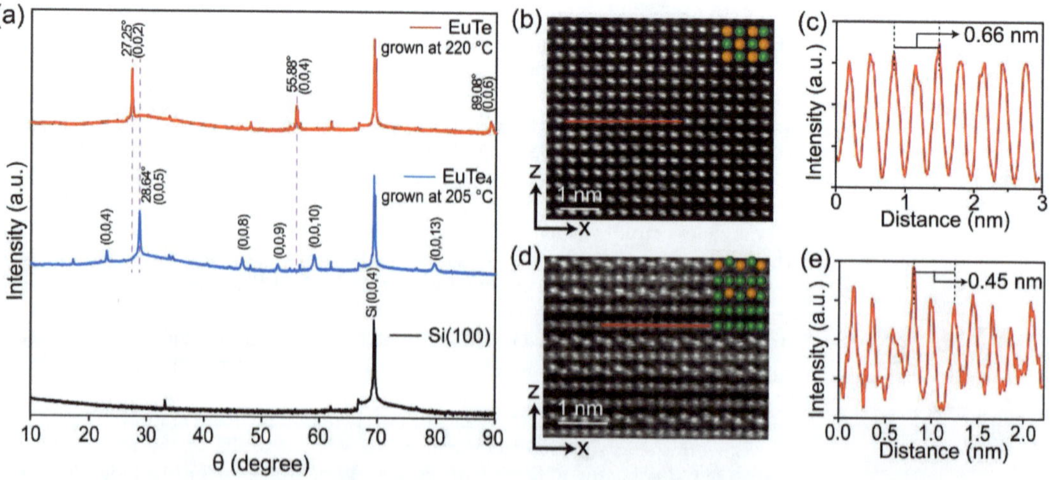

Figure 3. (**a**) XRD patterns of a Si(100) substrate (black curve), a ~20 nm $EuTe_4$ film (blue curve), and a ~20 nm EuTe film (red curve). The vertical purple dashed lines serve as guides to highlight the spectral differences between EuTe and $EuTe_4$. (**b**) STEM image (side-view) for a ~10 nm EuTe film, corresponding to the (010) plane of EuTe. (**c**) Intensity profile corresponding to the red solid line in (**b**). (**d**,**e**) Analogous to (**b**,**c**) for a ~10 nm $EuTe_4$ film.

Figure 3b displays the side-view STEM image of the EuTe film (grown at ~220 °C), where atoms form a tetragonal lattice. A schematic atomic arrangement of EuTe is depicted in the top-right corner of the STEM image. The intensity distribution curve, shown in Figure 3c, derived along the red solid line in Figure 3b, reveals an in-plane lattice constant of a = 0.66 nm for EuTe. This is in line with the value of 0.65 nm for EuTe grown on the PbTe buffer layer on $BaF_2(111)$ [13]. According to the lattice configurations depicted in Figure 2c,f and the derived in-plane lattice constant, we obtained a lattice mismatch of 21.55% between EuTe and the silicon substrate, which significantly surpasses the value of 2.10% between EuTe(111) and the buffer layer of PbTe on $BaF_2(111)$ [13]. The large lattice mismatch indicates a weak interfacial interaction between the substrate and the thin film, ensuring the high-quality growth of EuTe.

Figure 3d is a side-view STEM image of the $EuTe_4$ film (grown at ~205 °C), which displays a layered atomic structure comprised of EuTe-Te-EuTe-Te-Te. The corresponding intensity distribution curve in Figure 3e yields an in-plane lattice constant of a = 0.45 nm for $EuTe_4$, consistent with the value of 0.451 nm derived from bulk $EuTe_4$ XRD characterization [20]. This result gives a lattice mismatch of 17.13% between the Si substrate and $EuTe_4$ film.

3.2. XPS Differences in EuTe and $EuTe_4$

The elemental stoichiometry and valence states of EuTe and $EuTe_4$ films were investigated by in situ XPS, with all measurements performed at 300 K. Figure 4a illustrates the full-range XPS spectra for EuTe (upper section) and $EuTe_4$ (lower section). The two spectra exhibit significant differences, with the signal intensity of Eu $3d_{3/2}$ and Eu $3d_{5/2}$ orbitals in EuTe being notably higher than that in $EuTe_4$, indicating a higher concentration of Eu in EuTe. To study the ratio of Te to Eu in EuTe and $EuTe_4$, we present a detailed scan of Eu $3d_{3/2}$, Eu $3d_{5/2}$, Te $3d_{3/2}$, and Te $3d_{5/2}$ orbitals in Figure 4b. The relative peak areas of each peak after background subtraction are listed in the left half of Table 1, where the peak area of the Te $3d_{5/2}$ orbital was normalized as unit one. It shows that the peak area ratios of the $3d_{3/2}$ to $3d_{5/2}$ orbitals for each element in both the EuTe and $EuTe_4$ closely match a 2:3 distribution, which is in line with the characteristics of p-orbital electrons in

XPS spectra, indicating our treatment of peak areas is accurate. The right half of Table 1 lists the comparative area ratios of Eu's individual *3d* orbital to those of Te *3d* orbitals. We further calculated the quotient of this ratio in EuTe relative to that in $EuTe_4$, as shown in Table 2. This result provides a representation of the relative Te content in $EuTe_4$ compared to EuTe. The resultant value ranges from approximately 3.79 to 3.93, closely approximating 4, which implies the Te content in the $EuTe_4$ is about four times compared to that in EuTe. The error primarily stems from the different photon–electron cross-sections of the different Te elements in EuTe and $EuTe_4$.

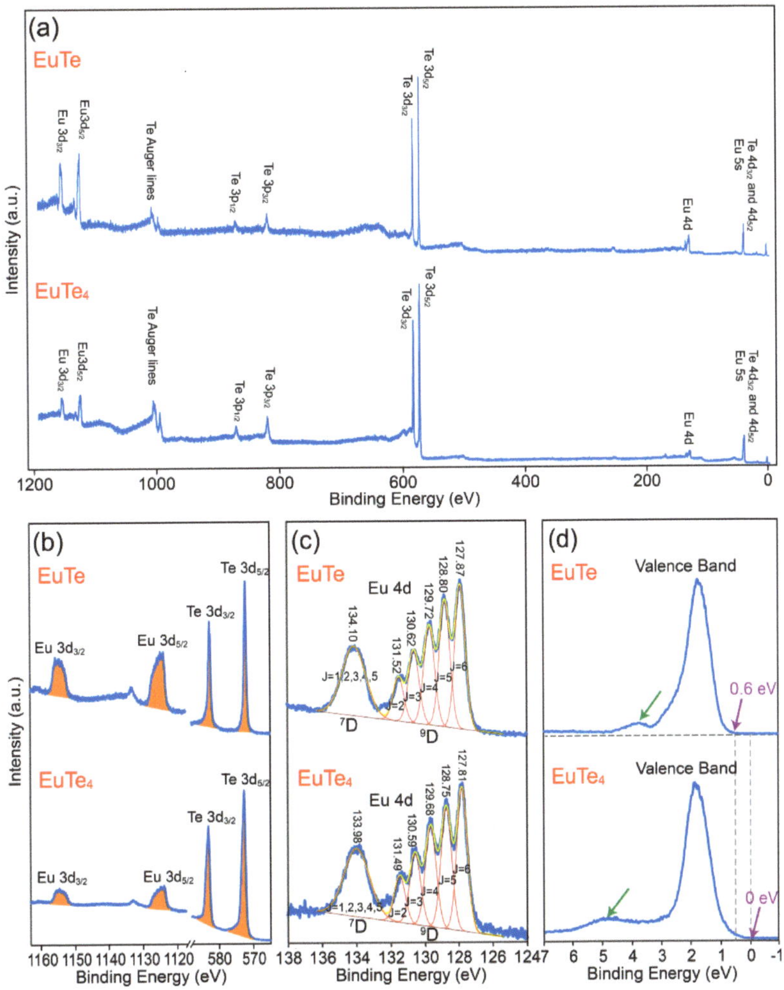

Figure 4. XPS spectra for ~10 nm EuTe (**top** of each panel) and $EuTe_4$ (**bottom** of each panel) films, including (**a**) the full range spectra, (**b**) the spectra for Eu *3d* and Te *3d* orbitals, (**c**) the spectra for Eu 4d orbital, and (**d**) the spectra near Fermi level. The horizontal gray dashed line in (**d**) denotes the baseline of EuTe spectra, while the vertical gray dashed lines indicate the cutoff energy of EuTe and $EuTe_4$, respectively.

Table 1. Normalized peak areas of Eu and Te *3d* orbits and their comparative ratios.

	EuTe	$EuTe_4$		EuTe	$EuTe_4$
Eu $3d_{3/2}$	0.934	0.238	Te $3d_{3/2}$:Eu $3d_{3/2}$	0.718	2.719
Eu $3d_{5/2}$	1.423	0.362	Te $3d_{5/2}$:Eu $3d_{3/2}$	1.071	4.211
Te $3d_{3/2}$	0.671	0.646	Te $3d_{3/2}$:Eu $3d_{5/2}$	0.471	1.782
Te $3d_{5/2}$	1	1	Te $3d_{5/2}$:Eu $3d_{5/2}$	0.703	2.760

Table 2. Ratio of Te content in $EuTe_4$ to EuTe, based on the data in Table 1.

	Te $3d_{3/2}$:Eu $3d_{3/2}$	Te $3d_{5/2}$:Eu $3d_{3/2}$	Te $3d_{3/2}$:Eu $3d_{5/2}$	Te $3d_{5/2}$:Eu $3d_{5/2}$
$EuTe_4$:EuTe	3.787	3.933	3.783	3.928

Figure 4c displays the detailed scanning spectra along with the fitting curves of Eu *4d* orbitals in EuTe and $EuTe_4$, respectively, where the position of each peak is listed above the corresponding curve The leftmost peak of each Eu *4d* orbital comprises five orbitals $^7D_1$, $^7D_2$, $^7D_3$, $^7D_4$, and $^7D_5$, and the five peaks on the right represent the five orbitals $^9D_2$, $^9D_3$, $^9D_4$, $^9D_5$, and $^9D_6$ [36]. The *4d* orbital spectra in EuTe and $EuTe_4$ show similar peak shapes and positions within the experimental error range, further confirming that the valence state of Eu remains unchanged (+2 state) in EuTe and $EuTe_4$.

Meanwhile, we present the XPS spectra near the Fermi level in Figure 4d. The peak width and position of their respective leftmost peaks (highlighted by the green arrows) exhibit distinct differences. Specifically, this peak in $EuTe_4$ exhibits a broader width compared to that in EuTe, and its peak position shifts to a deeper binding energy. Moreover, the spectrum of EuTe cuts off at ~0.6 eV, indicating that EuTe is a semiconductor with its valence band maximum located ~0.6 eV below the Fermi level. This is consistent with the 2.26 eV band gap observed in EuTe on the $BaF_2(111)$ substrate [37]. In contrast, the spectrum for $EuTe_4$ stretches close to the Fermi level. From an energy band theory perspective, we can infer from the XPS spectrum that $EuTe_4$ exhibits a density of states in the vicinity of the Fermi level (from −0.2 to 0 eV). Considering that our XPS resolution is ~0.2 eV, this result suggests that $EuTe_4$ is either a small gap semiconductor or a metallic material. Previous ARPES results reported that $EuTe_4$ has a CDW gap of ~0.2 eV at the Fermi level [7,8,19]. This value aligns well with our XPS measurements, especially considering the XPS resolution of ~0.2 eV.

3.3. Physical Property Characterizations of EuTe and $EuTe_4$

We conducted ex situ transmittance spectroscopy on the two films, which is a widely used technique in probing the band gap of a material [37]. Figure 5a displays a photograph of EuTe and $EuTe_4$ films before transmittance spectroscopy characterization. Notably, the films exhibit distinctly different colors, where EuTe appears green and $EuTe_4$ is golden yellow, indicating the transmittance spectra of the two films are different.

Figure 5b,c present the transmittance spectra of EuTe and $EuTe_4$, respectively. The process of transmittance spectroscopy characterization involves two steps. We first measured the transmittance spectrum of the apparatus and silicon substrate, represented as $T_1 = T_{appa} \times T_{sub}$, which serves as a reference value. The result of T_1 is shown in Supplementary Material Part B. Then we measured the transmittance spectrum of the apparatus, silicon substrate, and film: $T_2 = T_{appa} \times T_{sub} \times T_{film}$. Dividing the two values yields the transmittance spectrum of the film: T_{film}. The band gap of our silicon substrate is approximately 1.1 eV, which corresponds to a wavenumber close to 10,000 cm^{-1}. As a result, the transmittance spectrum of silicon substrate drops sharply to zero beyond this value. This leads to an indeterminate form in the T_{film} expression, resembling 0/0. Consequently, the transmittance spectrum for our film diverges above ~10,000 cm^{-1}, making the data valid only for a wavenumber below this threshold.

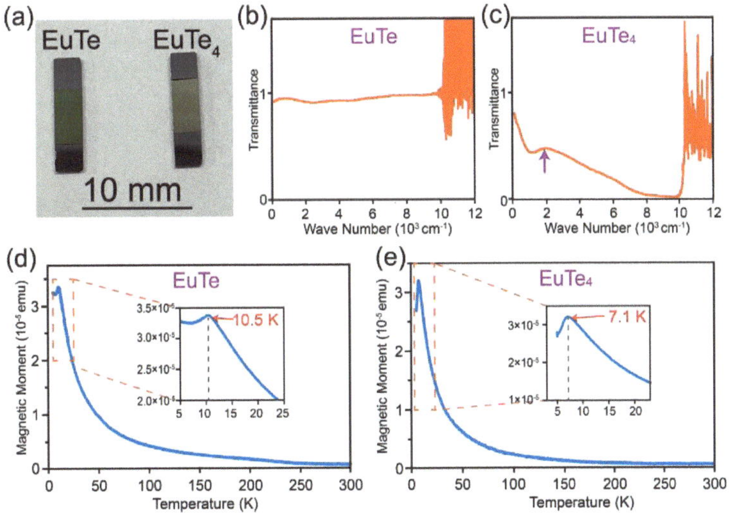

Figure 5. (**a**) Photograph of the epitaxial EuTe and $EuTe_4$ films. (**b**,**c**) Transmittance curves of ~20 nm thick EuTe and $EuTe_4$ film, respectively. (**d**,**e**) M−T curves of ~40 nm thick EuTe and $EuTe_4$ film, respectively. Insets in (**d**,**e**) are the corresponding magnified curves at low temperatures.

In Figure 5b, the transmittance spectrum for EuTe is flat and remains close to 1 within the experimentally accessible range, indicating that EuTe is a semiconductor with a band gap exceeding 1.1 eV. This is consistent with the earlier optical transmittance measurement on EuTe grown on the BaF_2 substrate, which identified a band gap of 2.26 eV [37].

In contrast, the transmittance curve for $EuTe_4$ in Figure 5c presents a distinct absorption edge around 1900 cm^{-1} (highlighted by the purple arrow), corresponding to an energy of approximately 0.23 eV. Previous ARPES experiments have demonstrated the presence of a gap in $EuTe_4$ induced by CDW at room temperature, with a size of ~0.2 eV [7,8,19]. This matches the energy of the absorption edge observed in our transmittance spectrum. Thus, this absorption edge is a manifestation of the CDW gap, specifically attributed to optical electron excitations across the CDW gap of ~0.23 eV.

Figure 5d,e are the magnetic moment versus temperature (M−T) curves of EuTe and $EuTe_4$. The shapes of the two curves are very similar, both exhibiting sharp peaks at low temperatures, which is a typical feature of antiferromagnetic material. Here, we magnified and plotted the details of these curves at low temperatures in the inset. Consequently, we can derive the Néel temperatures of EuTe and $EuTe_4$ to be 10.5 and 7.1 K, respectively, which are consistent with the 9.8 and 7.1 K reported in previous magnetic susceptibility measurements [18,20].

4. Conclusions

In conclusion, we have successfully synthesized high-quality EuTe and $EuTe_4$ thin films on Si(100) substrates. Our study improves the growth procedure of epitaxial EuTe films and fills the research gap in the synthesis of two-dimensional $EuTe_4$ films. We conducted a comprehensive study on the structural and energy spectra characterization of the two materials, confirming their antiferromagnetic nature. We also verified the semiconductive property of EuTe and found the CDW signature of $EuTe_4$ at room temperature. By tailoring the substrate temperature, we have achieved selective growth of these two materials, opening new possibilities for their physical property control. Our research on rare earth tellurides has enriched the library of 2D materials. The high-quality growth of the film paves the way for subsequent related research such as 2D magnetism and charge density waves, and also facilitates the exploration of potential applications in electronics.

Supplementary Materials: The following supporting information can be downloaded at: https://www.mdpi.com/article/10.3390/ma16227093/s1, Supplementary Material: A. Optimization of substrate temperature; B. More details on the transmittance spectroscopy.

Author Contributions: Conceptualization, F.Y. and Y.Z.; Methodology, F.Y., J.L., D.W. and Y.Z.; Validation, F.Y., J.L., D.W., G.W. and Y.Z.; Formal Analysis, F.Y. and Y.Z.; Investigation, F.Y., X.Q., J.Z., L.H. and B.Y.; Data Curation F.Y. and Y.Z.; Visualization F.Y. and Y.Z.; Writing—Original Draft Preparation, F.Y.; Writing—Review and Editing, G.W. and Y.Z.; Supervision, J.L., D.W., G.W. and Y.Z.; Project Administration, Y.Z.; Funding Acquisition, Y.Z. All authors have read and agreed to the published version of the manuscript.

Funding: This work is supported by the National Natural Science Foundation of China (No. 92165205), the Innovation Program for Quantum Science and Technology of China (No. 2021ZD0302803), the National Key Research and Development Program of China (No. 2018YFA0306800), and the Program of High-Level Entrepreneurial and Innovative Talents Introduction of Jiangsu Province, China.

Institutional Review Board Statement: Not applicable.

Informed Consent Statement: Not applicable.

Data Availability Statement: The data presented in this study are available on request from the corresponding author.

Acknowledgments: We are deeply grateful to Junhu Zhang from Nanjing University for their helpful discussions on XRD data. We thank Zhe Liu, Xiaoxiang Zhou and Yaomin Dai from Nanjing University for their assistance on the transmittance characterizations. We thank Qinghao Meng, Qichao Tian and Kaili Wang from Nanjing University for their assistance on the MBE growth of samples and transmittance characterizations.

Conflicts of Interest: The authors declare no competing interest.

References

1. Kogar, A.; Zong, A.; Dolgirev, P.E.; Shen, X.; Straquadine, J.; Bie, Y.-Q.; Wang, X.; Rohwer, T.; Tung, I.C.; Yang, Y.; et al. Light-induced charge density wave in $LaTe_3$. *Nat. Phys.* **2020**, *16*, 159–163. [CrossRef]
2. Brouet, V.; Yang, W.L.; Zhou, X.J.; Hussain, Z.; Ru, N.; Shin, K.Y.; Fisher, I.R.; Shen, Z.X. Fermi surface reconstruction in the CDW state of $CeTe_3$ observed by photoemission. *Phys. Rev. Lett.* **2004**, *93*, 126405. [CrossRef] [PubMed]
3. Lee, E.; Kim, D.H.; Kim, H.W.; Denlinger, J.D.; Kim, H.; Kim, J.; Kim, K.; Min, B.I.; Min, B.H.; Kwon, Y.S.; et al. The 7 $\times$ 1 Fermi surface reconstruction in a two-dimensional f -electron charge density wave system: $PrTe_3$. *Sci. Rep.* **2016**, *6*, 30318. [CrossRef] [PubMed]
4. Yumigeta, K.; Qin, Y.; Li, H.; Blei, M.; Attarde, Y.; Kopas, C.; Tongay, S. Advances in rare-earth tritelluride quantum materials: Structure, properties, and synthesis. *Adv. Sci.* **2021**, *8*, 2004762. [CrossRef] [PubMed]
5. Liu, J.S.; Huan, S.C.; Liu, Z.H.; Liu, W.L.; Liu, Z.T.; Lu, X.L.; Huang, Z.; Jiang, Z.C.; Wang, X.; Yu, N.; et al. Electronic structure of the high-mobility two-dimensional antiferromagnetic metal $GdTe_3$. *Phys. Rev. Mater.* **2020**, *4*, 114005. [CrossRef]
6. Lei, S.; Lin, J.; Jia, Y.; Gray, M.; Topp, A.; Farahi, G.; Klemenz, S.; Gao, T.; Rodolakis, F.; McChesney, J.L.; et al. High mobility in a van der Waals layered antiferromagnetic metal. *Sci. Adv.* **2020**, *6*, eaay6407. [CrossRef] [PubMed]
7. Lv, B.Q.; Zong, A.; Wu, D.; Rozhkov, A.V.; Fine, B.V.; Chen, S.-D.; Hashimoto, M.; Lu, D.-H.; Li, M.; Huang, Y.B.; et al. Unconventional hysteretic transition in a charge density wave. *Phys. Rev. Lett.* **2022**, *128*, 036401. [CrossRef]
8. Zhang, Q.Q.; Shi, Y.; Zhai, K.Y.; Zhao, W.X.; Du, X.; Zhou, J.S.; Gu, X.; Xu, R.Z.; Li, Y.D.; Guo, Y.F.; et al. Thermal hysteretic behavior and negative magnetoresistance in the charge density wave material $EuTe_4$. *Phys. Rev. B* **2023**, *107*, 115141. [CrossRef]
9. Van Houten, S. Magnetic interaction in EuS, EuSe, and EuTe. *Phys. Lett.* **1962**, *2*, 215–216. [CrossRef]
10. Oliveira, N.F.; Foner, S.; Shapira, Y.; Reed, T.B. EuTe. I. Magnetic behavior of insulating and conducting single crystals. *Phys. Rev. B* **1972**, *5*, 2634–2646. [CrossRef]
11. Silberstein, R.P.; Schmutz, L.E.; Tekippe, V.J.; Dresselhaus, M.S.; Aggarwal, R.L. Magnetic phase-dependent Raman scattering in EuSe and EuTe. *Solid State Commun.* **1976**, *18*, 1173–1177. [CrossRef]
12. Will, G.; Pickart, S.J.; Nathans, R. Antiferromagnetic structure of EuTe. *J. Phys. Chem. Solids* **1963**, *24*, 1679–1681. [CrossRef]
13. Springholz, G.; Bauer, G. Strain relaxation by coherent three-dimensional islanding in molecular-beam epitaxy of EuTe on PbTe(111). *Phys. Rev. B* **1993**, *48*, 10998–11009. [CrossRef] [PubMed]
14. Kępa, H.; Springholz, G.; Giebultowicz, T.M.; Goldman, K.I.; Majkrzak, C.F.; Kacman, P.; Blinowski, J.; Holl, S.; Krenn, H.; Bauer, G. Magnetic interactions in EuTe epitaxial layers and EuTe/PbTe superlattices. *Phys. Rev. B* **2003**, *68*, 024419. [CrossRef]
15. Bergomi, L.; Chen, J.J. Magnetic susceptibility of EuTe/PbTe Heisenberg antiferromagnetic superlattices: Experimental and theoretical studies. *Phys. Rev. B* **1997**, *56*, 3281–3289. [CrossRef]

16. Zinn, W. Microscopic studies of magnetic properties and interactions recent results on europium-monochalcogenides. *J. Magn. Magn. Mater.* **1976**, *3*, 23–36. [CrossRef]
17. Wachter, P. Chapter 19 Europium chalcogenides: EuO, EuS, EuSe and EuTe. In *Handbook on the Physics and Chemistry of Rare Earths*; Elsevier: Amsterdam, The Netherlands, 1979; Volume 2, pp. 507–574.
18. Scheer, E.; Wosnitza, J.; v. Löhneysen, H.; Kürsch, R.; Lang, M.; Steglich, F. Critical exponents of EuTe from specific-heat and thermal-expansion measurements. *J. Magn. Magn. Mater.* **1992**, *104–107*, 175–176. [CrossRef]
19. Zhang, C.; Wu, Q.-Y.; Yuan, Y.-H.; Zhang, X.; Liu, H.; Liu, Z.-T.; Zhang, H.-Y.; Song, J.-J.; Zhao, Y.-Z.; Wu, F.-Y.; et al. Angle-resolved photoemission spectroscopy study of charge density wave order in the layered semiconductor $EuTe_4$. *Phys. Rev. B* **2022**, *106*, L201108. [CrossRef]
20. Wu, D.; Liu, Q.M.; Chen, S.L.; Zhong, G.Y.; Su, J.; Shi, L.Y.; Tong, L.; Xu, G.; Gao, P.; Wang, N.L. Layered semiconductor $EuTe_4$ with charge density wave order in square tellurium sheets. *Phys. Rev. Mater.* **2019**, *3*, 024002. [CrossRef]
21. Rathore, R.; Pathak, A.; Gupta, M.K.; Mittal, R.; Kulkarni, R.; Thamizhavel, A.; Singhal, H.; Said, A.H.; Bansal, D. Evolution of static charge density wave order, amplitude mode dynamics, and suppression of Kohn anomalies at the hysteretic transition in $EuTe_4$. *Phys. Rev. B* **2023**, *107*, 024101. [CrossRef]
22. Springholz, G.; Bauer, G.; Schilcher, K. Epitaxial growth of EuTe on PbTe (111) influenced by strain-induced coherent 3D islanding. *MRS Online Proc. Libr.* **1993**, *317*, 137. [CrossRef]
23. Springholz, G.; Yuan, S.; Bauer, G.; Kriechbaum, M.; Krenn, H. EuTe/PbTe superlattices: MBE growth and optical characterization. *MRS Online Proc. Libr.* **1993**, *301*, 353. [CrossRef]
24. Frank, N.; Springholz, G.; Bauer, G.H. Strain induced islanding of EuTe epitaxial films observed by in-situ RHEED and STM investigations. *Mater. Sci. Forum* **1993**, *143–147*, 1623–1628. [CrossRef]
25. Chen, J.J.; Dresselhaus, G.; Dresselhaus, M.S.; Springholz, G.; Pichler, C.; Bauer, G. Magnetization studies of type-II antiferromagnetic EuTe/PbTe superlattices. *Phys. Rev. B* **1996**, *54*, 402–410. [CrossRef] [PubMed]
26. Heiss, W.; Prechtl, G.; Springholz, G. Giant tunability of exciton photoluminescence emission in antiferromagnetic EuTe. *Phys. Rev. B* **2001**, *63*, 165323. [CrossRef]
27. Ramachandran, V.; Smith, A.R.; Feenstra, R.M.; Greve, D.W. Temperature dependence of molecular beam epitaxy of GaN on SiC (0001). *J. Vac. Sci. Technol. A* **1999**, *17*, 1289–1293. [CrossRef]
28. Saito, Y.; Harima, H.; Kurimoto, E.; Yamaguchi, T.; Teraguchi, N.; Suzuki, A.; Araki, T.; Nanishi, Y. Growth temperature dependence of indium nitride crystalline quality grown by RF-MBE. *Phys. Status Solidi* **2002**, *234*, 796–800. [CrossRef]
29. Zhang, L.; Yang, T.; Feng, Y.P.; Wee, A.T.S.; Wang, Z. MBE-grown ultrathin $PtTe_2$ films and their layer-dependent electronic structures. *Nanoscale* **2022**, *14*, 7650–7658. [CrossRef] [PubMed]
30. Xu, H.; Han, D.; Bao, Y.; Cheng, F.; Ding, Z.; Tan, S.J.R.; Loh, K.P. Observation of gap opening in 1T′ phase MoS_2 nanocrystals. *Nano Lett.* **2018**, *18*, 5085–5090. [CrossRef] [PubMed]
31. Chen, P.; Pai, W.W.; Chan, Y.H.; Sun, W.L.; Xu, C.Z.; Lin, D.S.; Chou, M.Y.; Fedorov, A.V.; Chiang, T.C. Large quantum-spin-Hall gap in single-layer 1T′ WSe_2. *Nat. Commun.* **2018**, *9*, 2003. [CrossRef] [PubMed]
32. Feng, R.; Wang, W.; Bao, C.; Zhang, Z.; Wang, F.; Zhang, H.; Yao, J.; Xu, Y.; Yu, P.; Ji, S.-H.; et al. Selective control of phases and electronic structures of monolayer $TaTe_2$. *Adv. Mater.* **2023**, 2302297. [CrossRef]
33. Knall, J.; Sundgren, J.E.; Hansson, G.V.; Greene, J.E. Indium overlayers on clean Si (100) 2×1: Surface structure, nucleation, and growth. *Surf. Sci.* **1986**, *166*, 512–538. [CrossRef]
34. Cai, X.; Lu, Z.; Xu, Z.; Meng, F.; Zhang, Q.; Gu, L.; Feng, J.; Ji, S.-H.; Li, N.; Chen, X. Growth of (111)-orientated GdTe and TmTe thin films by van der Waals molecular beam epitaxy. *J. Phys. Chem. C* **2021**, *125*, 15465–15471. [CrossRef]
35. Koppensteiner, E.; Springholz, G.; Hamberger, P.; Bauer, G. Molecular beam epitaxy of PbTe/EuTe superlattices and their structural investigation by x-ray diffraction using reciprocal space mapping. *J. Appl. Phys.* **1993**, *74*, 6062–6071. [CrossRef]
36. Hüfner, S. Core levels and final states. *Photoelectron. Spectrosc. Princ. Appl.* **2003**, 61–107. [CrossRef]
37. Heiss, W.; Kirchschlager, R.; Springholz, G.; Chen, Z.; Debnath, M.; Oka, Y. Magnetic polaron induced near-band-gap luminescence in epitaxial EuTe. *Phys. Rev. B* **2004**, *70*, 035209. [CrossRef]

materials

Article

Hexagonal Boron Nitride as an Intermediate Layer for Gallium Nitride Epitaxial Growth in Near-Ultraviolet Light-Emitting Diodes

Ah-Hyun Park [1] and Tae-Hoon Seo [2,*]

1 R&D Center, Flyer, Daejeon 34141, Republic of Korea; ahpark14@gmail.com
2 Green Energy & Nano Technology R&D Group, Korea Institute of Industrial Technology, Gwangju 61012, Republic of Korea
* Correspondence: thseo@kitech.re.kr; Tel.: +82-62-600-6530

Abstract: We introduce the development of gallium nitride (GaN) layers by employing graphene and hexagonal boron nitride (h-BN) as intermediary substrates. This study demonstrated the successful growth of GaN with a uniformly smooth surface morphology on h-BN. In order to evaluate the crystallinity of GaN grown on h-BN, a comparison was conducted with GaN grown on a sapphire substrate. Photoluminescence spectroscopy and X-ray diffraction confirmed that the crystallinity of GaN deposited on h-BN was inferior to that of GaN grown on conventional GaN. To validate the practical applicability of the GaN layer grown on h-BN, we subsequently grew an NUV-LED structure and fabricated a device that operated well in optoelectrical performance experiments. Our findings validate the potential usefulness of h-BN to be a substrate in the direct growth of a GaN layer.

Keywords: hexagonal boron nitride; GaN; graphene; near-ultraviolet light-emitting diode

Citation: Park, A.-H.; Seo, T.-H. Hexagonal Boron Nitride as an Intermediate Layer for Gallium Nitride Epitaxial Growth in Near-Ultraviolet Light-Emitting Diodes. *Materials* **2023**, *16*, 7216. https://doi.org/10.3390/ma16227216

Academic Editors: Sebastiano Vasi and Ulderico Wanderlingh

Received: 30 October 2023
Revised: 15 November 2023
Accepted: 16 November 2023
Published: 17 November 2023

1. Introduction

The epitaxial development of GaN-based light-emitting diodes (LEDs) has brought about a significant transformation in the solid-state lighting sector [1–4]. GaN-based near-ultraviolet LEDs (NUV-LEDs) emitting light in the 370–400 nm range are employed as effective excitation sources for inorganic and organic luminescent materials used in white light production [5,6] and have been extensively used in various applications over the past few decades due to significant advancements in device efficiency, durability, and stability within the field of technology [7,8]. While the latter two qualities are, to a large extent, inherent to the material's properties, the device efficiency, i.e., the sum of internal and external quantum efficiencies, is principally decided by the structural and optical quality of the active layers and the device configuration. To optimize the internal quantum efficiency to its maximum extent, the most ideal approach is to grow the active layers on a native GaN substrate. GaN-based NUV-LEDs that are constructed via hetero-epitaxial growth on c-plane sapphire, SiC, or Si, as native GaN substrates, are prohibitively expensive. Among several substrates, the c-plane sapphire has many merits, including high-temperature resistance, preservation of the hexagonal crystal structure, and cost-effectiveness. However, the direct growth of an NUV-LED on a sapphire substrate leads to a low-quality layer and the eventual deterioration of device performance because of significant differences in the fundamental properties of the GaN layer and the sapphire substrate, such as lattice constants and thermal expansion coefficients, resulting in highly dense threading dislocations (TDs) [9,10]. It is noteworthy that this mismatch problem is a primary hurdle that needs to be addressed to achieve highly efficient NUV-LEDs.

Recently, the epitaxial growth of sp^3-bonded group-III-nitrides on sp^2-bonded two-dimensional (2D) materials, such as hexagonal boron nitride (h-BN) or graphene, has garnered significant attention due to its remarkable physical characteristics, which include high thermal conductivities, chemical and thermal stability, and mechanical flexibility [11–21]. In this case, the 2D material serves a dual purpose, functioning both as a buffer layer featuring a hexagonal in-plane lattice arrangement and as a release layer for mechanically exfoliating the layer. This approach has the potential to enable the production of large-scale flexible III-nitride devices. Achieving the direct epitaxial growth of a GaN layer on 2D materials poses challenges, primarily stemming from the low surface energies of 2D materials due to the absence of dangling bonds along the c-plane [11–19]. To resolve this issue, several pioneering studies have demonstrated that the epitaxial growth of a GaN layer on graphene or h-BN can be realized by integrating additional processing that forms dangling bonds or by adding layers, such as zinc oxide nano-walls [11,12], AlN [13,15,18], carbon nanotubes [17], and nanorods [20]. In addition, Wu et al. introduced the growth of an AlGaN-based deep-ultraviolet LED structure on oxygen-plasma-treated h-BN/Al_2O_3 [22]. Liu et al. also reported the interfacial bonding behavior and nucleation phenomena of a GaN layer grown on an activated h-BN/sapphire substrate [19]. Although the growth of a GaN layer on graphene or h-BN has been reported, differences in the nucleation behavior of a GaN layer grown on graphene and those grown on h-BN have seldom been reported.

In this study, we utilized graphene and h-BN as intermediary substrates to analyze disparities in the nucleation characteristics of GaN layers grown on them. It was successfully deposited on h-BN/sapphire substrates with the aid of a traditional low-temperature GaN buffer layer. Note that although a flat GaN layer has been successfully grown on both graphene and h-BN by pioneering works [11,15,19], growing GaN on graphene is not as easily accessible as that on h-BN. We also grew GaN-based NUV-LEDs based on these planar GaN layers on the h-BN/sapphire substrate and compared their performance with that of conventional sapphire-grown NUV-LEDs.

2. Materials and Methods

2.1. Synthesis of h-BN

A large-scale h-BN layer investigated in this work was synthesized on 35 μm thick Cu foils (sourced from Nippon Mining) using low-pressure chemical vapor deposition (LPCVD). The copper foil's surface was made flat via an electrochemical polishing (ECP) process, conducted in a solution containing phosphoric acid and water for 10 min at 1.8 V. In this process, a Cu plate was employed as the cathode, whereas a 100 × 100 mm copper foil served as the working electrode. Borazine ($B_3N_3H_6$), as a precursor to h-BN, was stored in a bubbler-equipped canister in a chiller at −10 °C. Following this, the copper foil that had been treated with ECP was placed in the middle position of a quartz tube and heated using a split-tube furnace. Meanwhile, the quartz tube was pumped down to 0.018 torr, and H_2 gas was flown through the reactor at 15 sccm during the temperature ramp-up up to 1040 °C. The annealing step was carried out at 1040 °C for 60 min under flowing H_2 gas (15 sccm) at a specified low pressure. The h-BN was synthesized by introducing a mixture of borazine (0.3 sccm) and H_2 (70 sccm) at 1040 °C for 90 s at a pressure of 5×10^{-3} torr. These conditions are ideal for achieving high-quality uniform h-BN. The sample was subsequently rapidly cooled to room temperature in a hydrogen environment, after which the h-BN was transferred onto various substrates, including sapphire, glass, and SiO_2/Si, using a technique similar to that used to transfer graphene, in order to apply various substrates and investigate the structural properties of h-BN. Additional information is provided in our previous report [23].

2.2. Synthesis of Graphene

The graphene that was applied in our study was produced on the ECP-treated copper foil by the use of LPCVD. The copper foil was introduced into a quartz tube with a diameter of 4 inches and placed under an ongoing temperature increase of up to 1030 °C for a duration of 60 min. This process was carried out in the environment of H_2 gas flowing at a rate of 15 sccm using a split-tube furnace. At the same time, the chemical vapor deposition (CVD) chamber conducted a vacuum process, lowering the pressure to 0.072 torr. Following that, copper foils underwent an annealing process for 60 min. To obtain high-quality graphene, we employed the two-step growth method. In the initial stages, the attainment of large-size graphene domains was successfully achieved on a copper foil via the reduction in nucleation sites in an ambient atmosphere consisting of a mixture of CH_4 (5 sccm) and H_2 (100 sccm) gases. This process was carried out for a duration of 60 s at a temperature of 1030 °C. In the next step, the continuous graphene surface was accomplished by enlarging the flow rate of CH_4 (13 sccm) and increasing the growth time to 8 min. This was carried out while keeping the H_2 flow rate and temperature identical. The chamber was finally cooled to room temperature by injecting Ar gas (500 sccm). Further details are presented in a previous report [24].

2.3. 2D Materials (h-BN or Graphene) Transfer

In order to protect and transfer the 2D materials onto the desired substrate, polymethylmethacrylate (PMMA) was placed onto the surface of 2D materials using a spin-coating process. The spin-coating was carried out at a speed of 4000 rpm for 50 s. Prior to the 2D materials transfer, the unintentional formation of 2D materials on the opposite side of the copper foil was eliminated via the utilization of O_2 plasma etching. Subsequently, the copper foil with PMMA-covered 2D materials was immersed in a solution containing 0.1 M ammonium persulfate $[(NH_4)_2S_2O_8]$ for 4 h, allowing the process of etching the copper foil. Then, PMMA/2D materials were transferred onto various substrates, such as the 300 nm SiO_2/Si substrate, the c-plane sapphire, glass, and Cu mesh TEM grid, to investigate the properties of 2D materials and to grow GaN. The PMMA was removed using acetone, and the sample was annealed for 2 h at 500 °C in an Ar/H_2 gas mixture to eliminate any remaining PMMA residues.

2.4. Growth of un-Doped GaN on an h-BN/Sapphire Substrate

A GaN epilayer was formed using metal–organic chemical vapor deposition (MOCVD) on an h-BN/sapphire substrate under identical growth conditions to those used to directly grow the GaN layer on a sapphire substrate. A 25 nm thick GaN buffer layer was applied to a sapphire substrate at 560 °C for 100 s under a growth pressure of 400 mbar as part of the conventional GaN growth process. Following this, a 3.7 μm thick undoped GaN layer was grown for 2 h at 1130 °C and 100 mbar.

2.5. Growth and Fabrication of the NUV-LED Structure

The LED configuration included an undoped GaN layer, a Si-doped n-type GaN layer, five sets of InGaN/AlGaN multi-quantum wells (MQWs), a Mg-doped p-AlGaN electron-blocking layer (EBL), and a Mg-doped p-type GaN layer. A 2 μm thick Si-doped n-type GaN layer was deposited on the undoped GaN layer on the h-BN/sapphire substrate at 1100 °C and 400 mbar for 60 min. Following this, five sets of $In_{0.04}Ga_{0.96}N$ quantum wells and $Al_{0.08}Ga_{0.92}N$ barrier layers, each with thicknesses of 3 and 12 nanometers, were grown at 720 °C and 810 °C, respectively, to serve as the active layers. Then, a 25 nm thick Mg-doped p-$Al_{0.25}Ga_{0.75}N$ electron-blocking layer (EBL) and a 100 nm thick p-type GaN contact layer were deposited at 1040 °C. To activate the Mg dopants, rapid thermal annealing was performed in a nitrogen (N_2) atmosphere at 940 °C for 40 s. The *p*-type GaN layer was calculated to have a hole concentration of approximately 10^{16} cm^{-3}. Figure 1 illustrates a schematic representation of an InGaN/AlGaN NUV-LED wafer developed on an h-BN/sapphire substrate. Further details are provided in the report of Seo et al. [25]. After LED wafer growth, individual LEDs, each with a chip dimension of 350 × 350 μm, were manufactured by utilizing an inductively coupled plasma etcher (ICP) with Cl_2/BCl_3/Ar gases to delineate the mesa region until the n-type GaN layer was revealed for contact with the n-electrode. Using an electron beam evaporator, a 200 nm thick layer of indium tin oxide (ITO) was then applied as a transparent current-spreading electrode on the *p*-type GaN layer. Finally, 50 nm thick Cr and 250 nm thick Au layers were deposited onto both the *n*-type GaN and ITO layers to serve as the *n*- and *p*-electrodes, employing an electron beam evaporator.

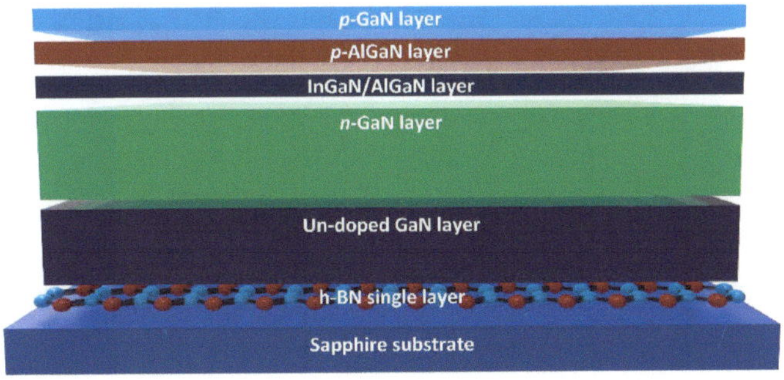

Figure 1. Schematic diagram of InGaN/AlGaN NUV-LED wafer grown on h-BN/sapphire substrate.

2.6. Formation of Patterned 2D Materials

The CVD-grown 2D materials were transferred onto a c-plane sapphire substrate to fabricate patterned 2D materials. Subsequently, the designated area of 3×3 µm was coated with a photoresist (PR), acting as a protective mask against the etchant. The 2D materials were then patterned by exposing them to an ICP using O_2 plasma. Finally, the PR was removed using acetone.

2.7. Characterization

Field-emission scanning electron microscopy (SEM) was employed to examine the surface structures of h-BN on copper foil, the initial phase of the GaN buffer layer on the h-BN/sapphire substrate, and the subsequent undoped GaN layer grown on these substrates. X-ray photoelectron spectroscopy (XPS) was conducted with the use of monochromatic aluminum K-alpha X-rays, utilizing equipment from Thermo Fisher Scientific. High-resolution transmission electron microscopy (HRTEM) was performed using an FEI TITAN G2 instrument equipped with an image Cs corrector and a monochromator provided by Thermo Fisher Scientific in Waltham, MA, USA. The equipment was operated at an optimized accelerator voltage of 60 kV to avoid damaging the h-BN. The crystallinities of the GaN layers, both with and without h-BN, were assessed via X-ray diffraction (XRD). For a comparative evaluation, samples were scanned at 2°/min in the 20–80° range. Photoluminescence (PL) spectroscopy, excited with the 325 nm line of a He–Cd laser, was employed to investigate the crystalline quality and residual strains in GaN layers cultivated on both sapphire and h-BN/sapphire substrates. Current–voltage (I-V) plots were constructed using a probe-station system, and electroluminescence (EL) experiments were performed.

3. Results and Discussion

In Figure 2a, one can observe an SEM image of h-BN production on copper foil via the ECP process. The h-BN completely covered the copper foil within 90 s of growth and exhibited a seamless two-dimensional nanosheet structure. However, wrinkles are commonly observed as they alleviate thermal stress; they may have originated through defect nucleation on the step margins of copper terraces during quenching, and their presence provides indirect evidence for the successful growth of continuous h-BN [26]. The h-BN thickness was determined using HRTEM, the results of which are depicted in Figure 2b, which verified that the h-BN synthesized in this study is predominantly monolayer. We acquired B and N core-level XPS spectra of the h-BN monolayer on the copper foil to confirm the growth of the h-BM monolayer and to quantify atomic concentrations. Atomic concentrations were determined from the intensity of the core-level photoemission for each element normalized by the atomic sensitivity factor at the photon emission energy. The B-to-N atomic ratio was calculated to be 1.02:0.98, which is close to 1:1. All XPS spectra were Lorentzian-fitted with multiple peaks. Figure 2c,d show B 1s and N 1s peaks at 189.2 and 396.8 eV, respectively, consistent with the binding energies reported for h-BN [27,28]. In addition, Raman spectroscopy was employed, which has become a crucial technique for characterizing and exploring two-dimensional materials to conduct a more in-depth analysis of the crystal structure of the h-BN monolayer. In Figure 2e, the Raman spectrum of the h-BN transferred onto the SiO2/Si substrate exhibits a peak at 1369.3 cm^{-1}, which is indicative of the E_{2g} lattice-vibration mode of h-BN in-plane oscillations. The peak at 1370 cm^{-1} corresponds to h-BN, whereas those associated with the cubic structure are observed at 1300 cm^{-1} (for the longitudinal optical vibrational mode) and 1065 cm^{-1} (for the transverse optical vibrational mode). Our sample did not exhibit the vibrational mode of cubic BN. Taking into account the number of h-BN layers, the E_{2g} mode of bulk h-BN is observed at approximately 1366 cm^{-1}, whereas the E_{2g} mode of the h-BN single layer is located between 1368 and 1370 cm^{-1}, and that of two-to-five-layer thick h-BN is observed between 1364 and 1367 cm^{-1} [29,30]. The identification of a peak at 1369.3 cm^{-1} in the h-BN indicates that it possesses a monolayer structure, which is in accordance with the HRTEM image of the sample presented in Figure 2b. The crystal size can be inferred from the full width at half maximum (FWHM) of the E_{2g} mode due to its correlation with the phonon vibration duration. The prepared h-BN has an FWHM value of approximately 26 cm^{-1}, which is comparable to the value of previously reported h-BN [31,32].

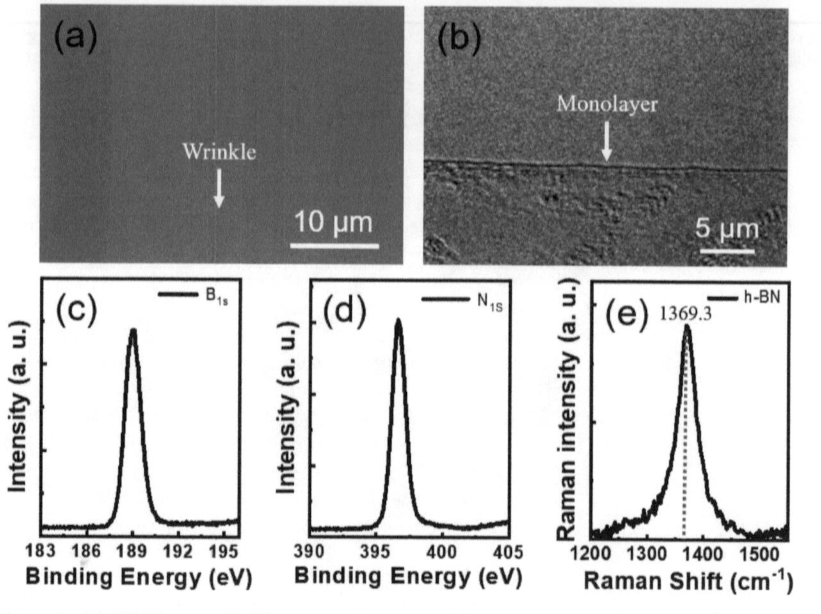

Figure 2. (**a**) SEM image of h-BN grown on copper foil; (**b**) HRTEM image of monolayer h-BN; (**c**,**d**) 1s core-level XPS spectra; and (**e**) Raman spectrum of the h-BN transferred onto the SiO_2/Si substrate.

The early stages of growth are crucial for achieving epitaxial layers and ensuring the superior quality of the resulting layer. To examine and compare GaN nucleation on h-BN, we acquired SEM images of the GaN buffer layer on a graphene/sapphire substrate and an h-BN/sapphire substrate, each with a 3 × 3 μm pattern size. This experiment enabled the initial GaN growth on sapphire, graphene, and h-BN to be compared. Figure 3a shows that dense and almost homogeneous nucleation occurred as the GaN buffer layer grew on sapphire. However, the GaN buffer layer grown on graphene exhibited irregular and low-density nucleation, which is attributable to the inherent non-reactivity of graphene and stems from its hexagonal arrangement of sp^2-bonded carbon atoms that lack dangling bonds [11]. The nucleated GaN exhibited an irregular three-dimensional growth pattern, leading to unevenly dispersed nucleation and resulting in a partially covered GaN surface after a growth period of 3 h (Figure 3c). In the case of h-BN (Figure 3b), the GaN buffer layer displays a consistent and organized GaN nucleation morphology, in contrast to the GaN buffer layer on graphene. However, GaN was less densely nucleated on h-BN than the GaN buffer layer grown on sapphire. The B-N bond types provide a reasonable explanation for the differences observed between graphene and h-BN. While the C-C- bonds in graphene are purely covalent and have evenly distributed electrons, the B-N bonds in the h-BN sheets display an alternating ionic/covalent nature due to the significant difference in the electronegativities of B and N [33,34]. We believe that electrically active N atoms attract the gallium atoms, which assists nucleation. First, growth begins only from GaN nucleation seeds that originate from N atoms or imperfections, such as point defects, wrinkles, and folds. These islands increase in size and spread progressively across the h-BN as growth progresses. The GaN crystallites eventually coalesce laterally to cover the entire surface. The GaN buffer layers do not cover the entire surface during the early growth following GaN deposition at low growth temperatures. This activity led to a noticeable transformation of the GaN surface, from a three-dimensional to a two-dimensional structure, on the h-BN during subsequent high-temperature growth, as shown by the SEM image in Figure 2d. The ultimate thickness of the GaN layer was ascertained to be 3.7 µm.

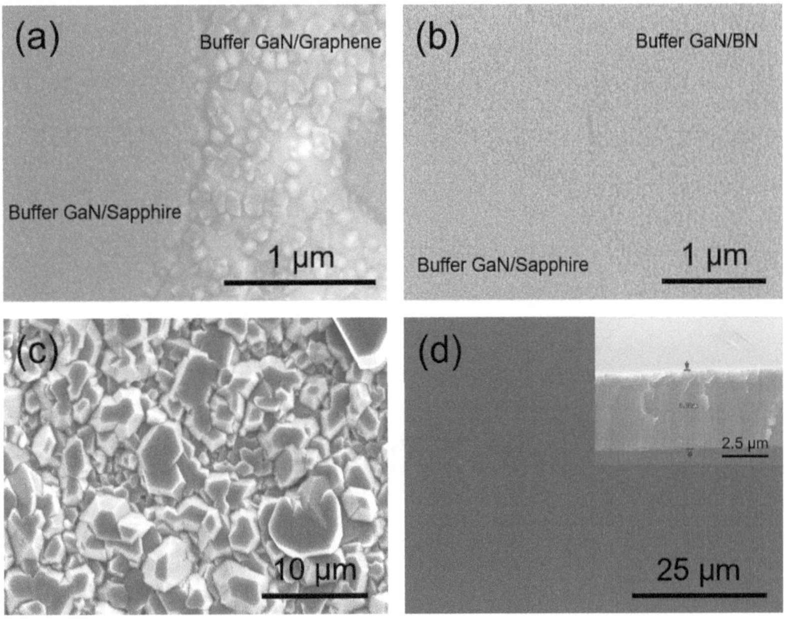

Figure 3. SEM images acquired during the initial step involved in the formation of the GaN buffer layer on (**a**) patterned graphene and (**b**) patterned h-BN. SEM images of undoped GaN formed on (**c**) graphene and (**d**) h-BN. The inset in panel (**d**) displays a cross-sectional SEM image.

We synthesized GaN on h-BN with a crack-free, mirror-like, and flat morphology. However, directly comparing a GaN layer on h-BN and its counterpart on graphene may not provide an entirely accurate result, considering that the GaN layer on graphene does not exhibit a typical two-dimensional morphology. We examined two samples to assess the crystallinity of the GaN layer on h-BN: GaN grown on h-BN and conventional GaN grown on a sapphire substrate. XRD omega-scanning is typically used to evaluate the crystallinity of a GaN layer. Figure 4 shows the XRD rocking curves for both the symmetric (002) and asymmetric (102) planes of GaN on sapphire and h-BN, which are commonly used to identify dislocation types. Specific lattice distortions are well known to influence the omega-scan FWHM values of crystal planes, such that the (002) plane is susceptible to screw or mixed dislocations, whereas the (102) plane is responsive to all types of dislocations, including pure-edge, screw, and mixed [17,35,36]. When compared to a conventional GaN layer grown on sapphire, the GaN grown on h-BN exhibits a slightly larger FWHM value for its (002) plane (308 vs. 268 arcsec), whereas the FWHM of the (102) plane is significantly larger (824 vs. 573 arcsec). Despite the slightly higher FWHM value for the (002) plane, the considerably higher value for the (102) plane implies the existence of more pure-edge dislocations in the GaN in h-BN than those on sapphire. The results presented herein suggest that the crystallinity of the GaN grown on h-BN is inferior to that grown on sapphire.

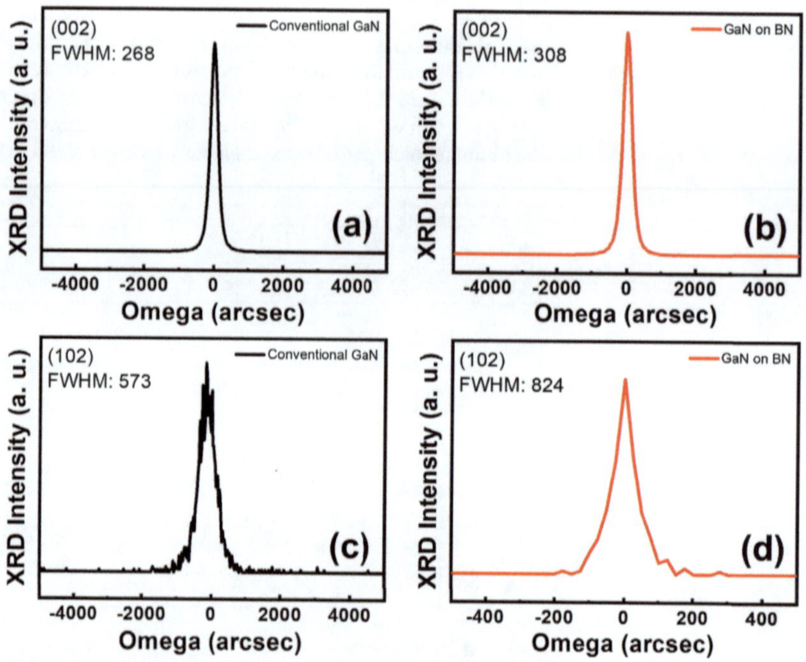

Figure 4. XRD omega rocking curves for the (**a**,**b**) symmetrical (002) and (**c**,**d**) asymmetrical (102) reflections of a GaN epilayer grown on sapphire and h-BN.

The impact of h-BN on the optical characteristics of the GaN layer was examined using PL spectroscopy, a rapid and nondestructive method, the results of which are depicted in Figure 5. The PL spectrum is notably influenced by the strain state and epilayer defects. Figure 5 depicts the room-temperature photoluminescence (PL) spectra of GaN grown separately on h-BN and on a sapphire substrate. Similar strong near-band-edge (NBE) emission peaks were observed at approximately 362 nm for both samples; however, GaN grown on h-BN exhibited an NBE emission peak intensity that was approximately 8% less intense than that grown on sapphire. The intensity of the NBE emission peak is generally acknowledged to be closely related to the defect density [37]. Although we successfully grew a smooth and crack-free GaN layer on h-BN, the GaN layer formed on h-BN appeared to contain additional defects, which is in good accord with the XRD results shown in Figure 4.

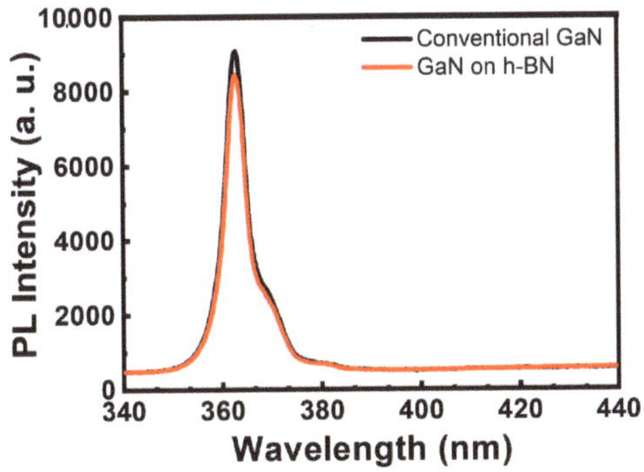

Figure 5. Room-temperature PL spectra of GaN epilayer formed on sapphire and h-BN.

We used a two-dimensional GaN layer on h-BN to fabricate NUV-LEDs, and LED devices with and without h-BN were fabricated to understand the real potential of the GaN layer grown on h-BN. We examined the effect of h-BN on the optoelectrical performance of the constructed NUV-LEDs. Figure 6a compares the I-V characteristics of an NUV-LED fabricated on h-BN with that fabricated on a sapphire substrate. The NUV-LEDs exhibited forward voltages of 3.25 V (on sapphire) and 3.37 V (on h-BN) at a driving current of 20 mA. The forward voltage of the NUV-LED on h-BN is slightly elevated compared to that on sapphire, which is rationalized by the higher number of defects and the greater internal strain in the GaN layer grown on h-BN. In general, TDs in LEDs serve as pathways for current leakage, thereby possibly increasing the forward voltage [38,39]; consequently, they serve as non-radiative recombination centers. The EL spectra of the NUV-LEDs fabricated on h-BN and sapphire substrates are shown in Figure 6b,c, respectively, which were obtained by varying the injection current in the 10–100 mA range. Both samples demonstrated adequate stability when tested at injection currents of up to 100 mA. The EL emission wavelength was observed to slightly redshift with increasing applied current (from 375 nm at 10 mA to 378 nm at 100 mA). The blue InGaN/GaN LED exhibited a lower emission peak energy than the theoretical value due to the quantum confinement Stark effect (QCSE). This shift originates from strong spontaneous polarization and internal piezoelectric fields associated with the high indium mole fraction. The peak position of the blue LED underwent a progressive shift toward higher energies with increasing excitation power, which is attributable to a higher photogenerated-carrier density leading to a lower QCSE and resulting in a blueshift in the emission peak as a consequence. The redshift behavior of the NUV-LEDs is a result of the relatively weak QCSE induced by a lower indium mole fraction [40]. The EL emission intensity of the NUV-LED on h-BN is 30% lower than that of the NUV-LED on the sapphire substrate at an injection current of 20 mA. The insets in Figure 6b,c present the electroluminescence (EL) emission images of the constructed devices at an injection current of 20 mA. The NUV-LED on the h-BN substrate is less bright than that on the sapphire substrate. Commercially advancing h-BN as a GaN growth substrate requires achieving a level of quality that exceeds that of the GaN layer grown on sapphire. We believe that a high-quality GaN layer on h-BN can be achieved via material optimization using active learning and lateral epitaxial overgrowth methods. Nevertheless, the findings of this study validate the potential utility of h-BN as a substrate.

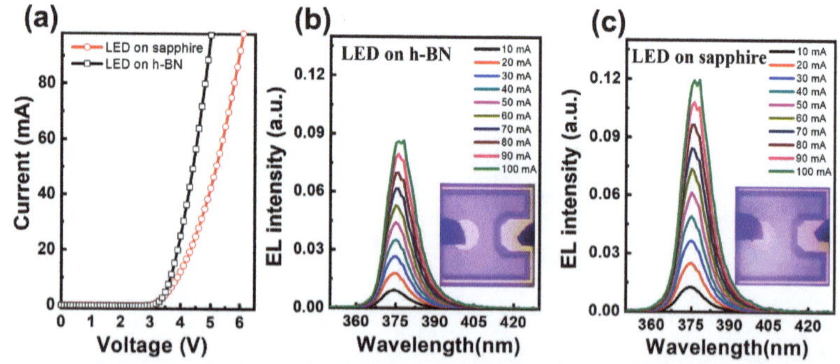

Figure 6. (**a**) I-V curves and (**b**,**c**) EL intensities with an injection current of fabricated NUV-LEDs on h-BN and sapphire. The insets display EL images at an applied current of 20 mA.

4. Conclusions

We employed graphene and h-BN as intermediate substrates to conduct a comparative analysis of the nucleation characteristics of GaN grown on these substrates. Although attempts to grow GaN on graphene failed to achieve a two-dimensional morphology, we successfully grew GaN on h-BN. The crystallinity of the GaN grown on h-BN was established via XRD and PL spectroscopy, with the results compared to those obtained using conventional GaN grown on sapphire. These findings reveal that the GaN produced on h-BN was less crystalline than that grown on sapphire despite the successful growth of crack-free GaN on h-BN. Planar GaN grown on h-BN forms a building block for the subsequent growth of a GaN-based NUV-LED; consequently, we fabricated LED devices. The optoelectrical performance of the manufactured NUV-LEDs was investigated to assess the effect of h-BN, with the LED grown on h-BN demonstrating excellent operational properties. This study demonstrated the potential applicability of h-BN as a substrate for GaN growth. We believe that the use of various growth strategies may enable h-BN substrates to surpass conventional sapphire substrates in terms of GaN growth quality.

Author Contributions: Conceptualization, A.-H.P. and T.-H.S.; methodology, A.-H.P.; validation, T.-H.S.; formal analysis, A.-H.P. and T.-H.S.; investigation, T.-H.S.; data curation, A.-H.P.; writing—original draft preparation, A.-H.P.; writing—review and editing, T.-H.S.; supervision, T.-H.S. All authors have read and agreed to the published version of the manuscript.

Funding: This research was carried out with the assistance of the Korea Institute of Industrial Technology as "Development of smart electric driving platform by eco-friendly power source in agricultural work environment" (KITECH JA-23-0008). This study received support from the Basic Science Research Program through the National Research Foundation of Korea (NRF), sponsored by the Ministry of Education, Science, and Technology (NRF-2022R1F1A1074422), as well as from the Korea Institute of Marine Science and Technology Promotion (KIMST), funded by the Ministry of Oceans and Fisheries, Korea (20200599).

Institutional Review Board Statement: Not applicable.

Informed Consent Statement: Not applicable.

Data Availability Statement: Data supporting the findings of this study is available from the corresponding author upon request.

Conflicts of Interest: The authors declare no conflict of interest.

References

1. Akasaki, I.; Amano, H.; Koide, Y.; Hiramatsu, K.; Sawaki, N. Effects of ain buffer layer on crystallographic structure and on electrical and optical properties of GaN and $Ga_{1-x}Al_xN$ ($0 < x \leq 0.4$) films grown on sapphire substrate by MOVPE. *J. Cryst. Growth* **1989**, *98*, 209–219.
2. Nakamura, S. GaN growth using GaN buffer layer. *Jpn. J. Appl. Phys.* **1991**, *30*, L1705–L1707. [CrossRef]

3. Nakamura, S.; Senoh, M.; Nagahama, S.; Iwasa, N.; Yamada, T.; Matsushita, T.; Kiyoku, H.; Sugimoto, Y.; Kozaki, T.; Umemoto, H.; et al. InGaN/GaN/AlGaN-based laser diodes with modulation-doped strained-layer superlattices grown on an epitaxially laterally overgrown GaN substrate. *Appl. Phys. Lett.* **1998**, *72*, 211–213. [CrossRef]
4. Nakamura, S.; Senoh, M.; Iwasa, N.; Negahama, S.-I. High-brightness InGaN blue, green and yellow light-emitting diodes with quantum well structures. *Jpn. J. Appl. Phys.* **1995**, *34*, L797–L799. [CrossRef]
5. Akasaki, I.; Amano, H. Crystal growth and conductivity control of group III nitride semiconductors and their application to short wavelength light emitters. *Jpn. J. Appl. Phys.* **1997**, *36 Pt 1*, 5393–5408. [CrossRef]
6. Hide, F.; Kozoday, P.; Denbaars, S.P.; Heeger, A.J. White light from InGaN/conjugated polymer hybrid light-emitting diodes. *Appl. Phys. Lett.* **1997**, *70*, 2664–2666. [CrossRef]
7. Templier, F. GaN-based emissive microdisplays: A very promising technology for compact, ultra-high brightness display systems. *J. Soc. Inf. Disp.* **2016**, *24*, 669–675. [CrossRef]
8. Lv, Q.; Liu, J.; Mo, C.; Zhang, J.; Wu, X.; Wu, Q.; Jiang, F. Realization of highly efficient InGaN green LEDs with sandwich-like multiple quantum well structure: Role of enhanced interwell carrier transport. *ACS Photonics* **2019**, *6*, 130–138. [CrossRef]
9. Jain, S.C.; Willander, M.; Narayan, J.; Overstraeten, R.V. III-nitrides: Growth, characterization, and properties. *J. Appl. Phys.* **2000**, *87*, 965–1006. [CrossRef]
10. Ning, X.J.; Chien, F.R.; Pirouz, P.; Yang, J.W.; Khan, M.A. Growth defects in GaN films on sapphire: The probable origin of threading dislocations. *J. Mater. Res.* **1996**, *11*, 580–592. [CrossRef]
11. Chung, K.; Lee, C.-H.; Yi, G.-C. Transferable GaN layers grown on ZnO-coated graphene layers for optoelectronic devices. *Science* **2010**, *330*, 655–657. [CrossRef] [PubMed]
12. Lee, C.H.; Kim, Y.J.; Hong, Y.J.; Jeon, S.R.; Bae, S.; Hong, B.H.; Yi, G.C. Flexible inorganic nanostructure light-emitting diodes fabricated on graphene films. *Adv. Mater.* **2011**, *23*, 4614–4619. [CrossRef] [PubMed]
13. Kobayashi, Y.; Kumakura, K.; Akasaka, T.; Makimoto, T. Layered boron nitride as a release layer for mechanical transfer of GaN-based devices. *Nature* **2012**, *484*, 223–227. [CrossRef] [PubMed]
14. Kong, W.; Li, H.; Qiao, K.; Kim, Y.; Lee, K.; Nie, Y.; Lee, D.; Osadchy, T.; Molnar, R.J.; Gaskill, D.K.; et al. Polarity governs atomic interaction through two-dimensional materials. *Nat. Mater.* **2018**, *17*, 999–1004. [CrossRef]
15. Gupta, P.; Rahman, A.A.; Hatui, N.; Goklhale, M.R.; Deshmukh, M.M.; Bhattacharya, A. MOVPE growth of semipolar III-nitride semiconductors on CVD graphene. *Cryst. Growth* **2013**, *372*, 105–108. [CrossRef]
16. Kim, J.; Bayram, C.; Park, H.; Cheng, C.-W.; Dimitrakopoulos, C.; Ott, J.A.; Reuter, K.B.; Bedell, S.W.; Sadana, D.K. Principle of direct van der Waals epitaxy of single-crystalline films on epitaxial graphene. *Nat. Comm.* **2014**, *5*, 4836. [CrossRef] [PubMed]
17. Seo, T.H.; Park, A.H.; Park, S.; Kim, Y.H.; Lee, G.H.; Kim, M.J.; Jeong, M.S.; Lee, Y.H.; Hahn, Y.; Suh, E.-K. Direct growth of GaN layer on carbon nanotube-graphene hybrid structure and its application for light emitting diodes. *Sci. Rep.* **2015**, *5*, 7747. [CrossRef]
18. Glavin, N.R.; Chabak, K.D.; Heller, E.R.; Moore, E.A.; Prusnick, T.A.; Maruyama, B.; Walker, D.E.; Dorsey, D.L., Jr.; Paduano, Q.; Snure, M. Flexible gallium nitride for high-performance, strainable radio-frequency devices. *Adv. Mater.* **2017**, *29*, 1701838. [CrossRef]
19. Liu, F.; Yu, Y.; Zhang, Y.; Rong, X.; Wang, T.; Zheng, X.; Sheng, B.; Yang, L.; Wei, J.; Wang, X.; et al. Hexagonal BN-assisted epitaxy of strain released GaN films for true green light-emitting diodes. *Adv. Sci.* **2020**, *7*, 2000917. [CrossRef]
20. Ren, F.; Liu, B.; Chen, Z.; Yin, Y.; Sun, J.; Zhang, S.; Jiang, B.; Liu, B.; Liu, Z.; Wang, J.; et al. Vna der Waals epitaxy of nearly single-crystalline nitride films on amorphous graphene-glass wafer. *Sci. Adv.* **2021**, *7*, eabf5011. [CrossRef]
21. Liu, B.; Chen, Q.; Chen, Z.; Yang, S.; Shan, J.; Liu, Z.; Yin, Y.; Ren, F.; Zhang, S.; Wang, R.; et al. Atomic mechanism of strain alleviation and dislocation reduction in highly mismatched remote heteroepitaxy using a graphene interlayer. *Nano Lett.* **1922**, *22*, 3364–3371. [CrossRef] [PubMed]
22. Wu, Q.; Guo, Y.; Sundaram, S.; Yan, J.; Zhang, L.; Wei, T.; Wei, X.; Wang, J.; Qugazzaden, A.; Li, J. Exfoliation of AlN film using two-dimensional multilayer hexagonal BN for deep-ultraviolet light-emitting diodes. *Appl. Phys. Express* **2019**, *12*, 015505. [CrossRef]
23. Lee, C.-H.; Choi, G.B.; Kim, E.M.; Lee, J.; Lee, J.; Moon, H.G.; Kim, M.J.; Kim, Y.A.; Seo, T.H. Gas barrier performance of hexagonal boron nitride monolayers grown on copper foils with electrochemical polishing. *Appl. Sci.* **2021**, *11*, 4599. [CrossRef]
24. Seo, T.H.; Lee, S.; Cho, H.; Chandramohan, S.; Suh, E.-K.; Lee, H.S.; Bae, S.K.; Kim, S.M.; Park, M.; Lee, J.K.; et al. Tailored CVD graphene coating as a transparent and flexible gas barrier. *Sci. Rep.* **2016**, *6*, 24143. [CrossRef] [PubMed]
25. Seo, T.H.; Kim, S.; Kim, M.J.; Kim, H.; Suh, E.-K. Compound Ag nanocluster-graphene electrodes as transparent and current spreading electrodes for improved light output power in near-ultraviolet light emitting diodes. *J. Phys. D Appl. Phys.* **2014**, *47*, 215103. [CrossRef]
26. Li, X.; Cai, W.; An, J.; Kim, S.; Nah, J.; Yang, D.; Piner, R.; Velamakanni, A.; Jung, I.; Tutuc, E.; et al. Large-area synthesis of high-quality and uniform graphene films on copper foils. *Science* **2009**, *324*, 1312–1314. [CrossRef] [PubMed]
27. Ismach, A.; Chou, H.; Ferrer, D.A.; Wu, Y.; McDonnell, S.; Floresca, H.C.; Covacevich, A.; Pope, C.; Piner, R.; Kim, M.J.; et al. Toward the controlled synthesis of hexagonal boron nitride films. *ACS Nano* **2012**, *6*, 6378–6385. [CrossRef]
28. Baazi, T.; Knystautas, E.J. Hexagonal boron nitride synthesis by nitrogen ion implantation of boron films. *Thin Solid Films* **1993**, *232*, 185–193. [CrossRef]

29. Gorbachev, R.V.; Riaz, I.; Nair, R.R.; Jalil, R.; Britnell, L.; Belle, B.D.; Hill, E.W.; Novoselov, K.S.; Watanabe, K.; Taniguchi, T.; et al. Hunting for monolayer boron nitride: Optical and Raman signatures. *Small* **2011**, *7*, 465–468. [CrossRef]
30. Reich, S.; Ferrari, A.C.; Arenal, R.; Loiseau, A.; Bello, I.; Robertson, J. Resonant Raman scattering in cubic and hexagonal boron nitride. *Phys. Rev. B* **2005**, *71*, 205201. [CrossRef]
31. Kim, K.K.; Hsu, A.; Jia, X.; Kim, S.M.; Shi, Y.; Hofmann, M.; Nezich, D.; Nieva, J.F.R.; Dresselhaus, M.; Palacios, T.; et al. Synthesis of monolayer hexagonal boron nitride on Cu foil using chemical vapor deposition. *Nano Lett.* **2012**, *12*, 161–166. [CrossRef] [PubMed]
32. Lee, K.H.; Shin, H.-J.; Lee, J.; Lee, I.-Y.; Kim, G.-H.; Choi, J.-Y.; Kim, S.-W. Large-scale synthesis of high-quality hexagonal boron nitride nanosheets for large-area graphene electronics. *Nano Lett.* **2012**, *12*, 714–718. [CrossRef]
33. Topsakal, M.; Akturk, E.; Ciraci, S. First-principles study of two- and one-dimensional honeycomb structures of boron nitride. *Phys. Rev. B Condens. Matter Mater. Phys.* **2009**, *79*, 115442. [CrossRef]
34. Hod, O. Graphite and hexagonal boron-nitride have the same interlayer distance. Why? *J. Chem. Theory Comput.* **2012**, *8*, 1360–1369. [CrossRef] [PubMed]
35. Zhang, L.; Li, X.; Shao, Y.; Yu, J.; Wu, Y.; Hao, X.; Yin, Z.; Dai, Y.; Tian, Y.; Huo, Q.; et al. Improving the quality of GaN crystals by using graphene or hexagonal boron nitride nanosheets substrate. *ACS Appl. Mater. Interfaces* **2015**, *7*, 4504–4510. [CrossRef] [PubMed]
36. Shih, H.-Y.; Shiojiri, M.; Chen, C.-H.; Yu, S.-F.; Ko, C.-T.; Yang, J.-R.; Lin, R.-B.; Chen, M.-J. Ultralow threading dislocation density in GaN epilayer on near-strain-free GaN compliant buffer layer and its applications in hetero-epitaxial LEDs. *Sci. Rep.* **2015**, *5*, 13671. [CrossRef] [PubMed]
37. Macht, L.; Weyher, J.L.; Grzegorczyk, A.; Larsen, P.K. Statistical photoluminescence of dislocations and associated defects in heteroepitaxial GaN grown by metal organic chemical vapor deposition. *Phys. Rev. B* **2005**, *71*, 073309. [CrossRef]
38. Lin, D.W.; Lin, C.C.; Chiu, C.H.; Lee, C.Y.; Yang, Y.Y.; Li, Z.Y.; Lai, W.C.; Lu, T.C.; Kuo, H.C.; Wang, S.C. GaN-based LEDs grown on HVPE growth high crystalline quality thick GaN template. *J. Electrochem. Soc.* **2011**, *158*, H1103–H1106. [CrossRef]
39. Han, N.; Park, Y.J.; Han, M.; Ryu, B.D.; Ko, K.B.; Chandramohan, S.; Choi, C.-J.; Cuong, T.V.; Hong, C.-H. Threading dislocation reduction in epitaxial GaN using V-groove patterned sapphire substrate with embedded silica nanospheres. *Mater. Lett.* **2014**, *123*, 97–100. [CrossRef]
40. Kuokstis, E.; Yang, J.W.; Simin, G.; Khan, M.A.; Gaska, R.; Shur, M.S. Two mechanisms of blueshift of edge emission in InGaN-based epilayers and multiple quantum wells. *Appl. Phys. Lett.* **2022**, *80*, 977–979. [CrossRef]

materials

MDPI

Article

Theoretical Study of the Phonon and Electrical Conductivity Properties of Pure and Sr-Doped $LaMnO_3$ Thin Films

Angel T. Apostolov [1], Iliana N. Apostolova [2] and Julia Mihailowa Wesselinowa [3,*]

1 University of Architecture, Civil Engineering and Geodesy, 1046 Sofia, Bulgaria; angelapos@abv.bg
2 University of Forestry, 1756 Sofia, Bulgaria; inaapos@abv.bg
3 Faculty of Physics, Sofia University "St. Kliment Ohridski", J. Bouchier Blvd. 5, 1164 Sofia, Bulgaria
* Correspondence: julia@phys.uni-sofia.bg

Abstract: The film thickness, temperature, substrate and doping dependence of the phonon energy ω and damping γ, as well as the electrical conductivity, of pure and Sr-doped $LaMnO_3$ thin films near the phase transition temperature T_N are investigated using a microscopic model and the Green's function technique. Due to the strong spin–phonon interaction, there appears a kink at T_N in the temperature dependence of $\omega(T)$ and $\gamma(T)$. The softening and hardening of the ω = 495 cm^{-1} (A_{1g}) and ω = 614 cm^{-1} (B_{2g}) modes is explained by the different sign of the anharmonic spin–phonon interaction constant R. The damping increases with T for both cases because it is proportional to R^2. ω decreases whereas γ increases with an increasing Sr concentration. This is due to the strain caused by the difference between the ionic radii of the La and Sr ions. The film thickness dependence is also considered. ω and γ increase strongly with the decreasing film thickness. The electrical conductivity is enhanced after the doping of the LMO thin films with Sr, which could be used for energy storage applications. The observed results are in good qualitative agreement with the experimental data.

Keywords: Sr-doped $LaMnO_3$ thin films; phonon energy and damping; electrical conductivity; microscopic model; Green's function theory

Citation: Apostolov, A.T.; Apostolova, I.N.; Wesselinowa, J.M. Theoretical Study of the Phonon and Electrical Conductivity Properties of Pure and Sr-Doped $LaMnO_3$ Thin Films. *Materials* **2024**, *17*, 1995. https://doi.org/10.3390/ma17091995

Academic Editors: Daniel N. Blaschke and Carlos Frontera

Received: 3 March 2024
Revised: 18 April 2024
Accepted: 23 April 2024
Published: 25 April 2024

1. Introduction

Ion-doped manganites with the general formula $R_{1-x}A_xMnO_3$ (R is a rare-earth element; A = Ca, Sr, Ba or Pb) and a perovskite structure possess interesting physical properties. Recently, Koriba et al. [1] have used density functional theory (DFT) in order to study the structural, electronic, magnetic and mechanical properties of $LaMnO_3$ (LMO) in its orthorhombic, cubic and rhombohedral phases. The properties of LMO have been studied intensively in the last few years since it was found that the partial substitution of La by Ca, Sr or Ba results in structural changes and the occurrence of colossal magnetoresistance near the temperatures of the spin ordering of Mn ions [2]. Hess et al. [3] have studied 20% (Ca, Sr, Ba)-doped LMO using a DFT-based defect chemistry model. Sr substitution at the A-site of LMO nanoparticles could be used for energy applications because it shows a higher current value and higher conductivity, as described by Gupta et al. [4].

LMO undergoes a phase transition from the low-temperature A-type antiferromagnetic phase to the paramagnetic phase at ~140 K and to the orbital disordered phase above 780 K. Let us emphasize that LMO thin films show ferromagnetic properties [5]. It must also be mentioned that antiferromagnetic nanostructures exhibit large magnetization and coercive fields below the phase transition temperature T_N compared to their bulk materials, which is due to the uncompensated surface spins [6]. By doping with Sr, bulk LMO becomes also ferromagnetic [7].

Although it is expected that structural changes and magnetic ordering will also influence the phonon spectra in Sr-doped LMO, these have not been sufficiently theoretically investigated. The changes in the lattice parameters with temperature manifest in a shift in the line position and intensity, as well as in the line width of the Raman peaks. Moreover, Raman spectroscopy is a useful method to study phase transitions. At 80 K, the lines of the A_g symmetry are observed at 148, 210, 267, 295, 452 and 495 cm^{-1} and those of the B_{1g} symmetry at 197, 313, 436 and 606 cm^{-1} [8]. The Raman spectra of Sr-doped bulk LMO manganites are studied in [9–11], whereas those of pure and Sr-doped LMO thin films and nanoparticles are investigated in [12–18]. Recently, Helton et al. [19] have studied the damping and softening of transverse acoustic phonons in $La_{0.7}Sr_{0.3}MnO_3$. The doping dependence of the phonon frequencies ω in Sr-doped LMO has been investigated experimentally by many authors [10,11,20,21]. It is observed that ω decreases when increasing the Sr concentration.

It must be mentioned that most theoretical works consider the magnetic properties of LMO. Wdowik et al. [22] calculated the vibrational dynamics of the undoped LMO from first principles by DFT. Talati et al. [23] and Rini et al. [24] investigated the phonon dispersion curves of LMO by using a lattice dynamical simulation method and an interatomic shell model potential, respectively.

The aim of the present paper is to use a microscopic model and the Green's function theory to study the phonon energy and damping, as well as the electrical conductivity, of Sr-doped LMO thin films in dependence on the film thickness, temperature, substrate and doping concentration. To our knowledge, such studies have not yet been performed using this method or density functional theory.

2. The Model

LMO, in which Mn^{3+} is the only present high-spin magnetic ion with $S = 2$, is an antiferromagnetic compound (A-type) whose Neel temperature is $T_N = 140$ K. It shows also orbital ordering (C-type) below 780 K, which will be not considered here. The Jahn–Teller electronic ordering couples the Mn^{3+} spins within the basal planes with ferromagnetic coupling (superexchange interaction). These planes are coupled one to another with antiferromagnetic coupling. The magnetic properties of ion-doped LMO with doping concentration x are described by the Heisenberg Hamiltonian:

$$H_{sp} = -\frac{1}{2}\sum_{i,j}(1-x)J_{ij}\mathbf{S}_i \cdot \mathbf{S}_j - \frac{1}{2}\sum_{i,j} xJ_{dij}\mathbf{S}_i \cdot \mathbf{S}_j - D_z\sum_i (S_i^z)^2, \tag{1}$$

where $\mathbf{S}_i$ and its z component S_i^z are spin operators for the localized Mn^{3+} spins at site i. J_{ij} stands for the spin exchange interaction between the nearest neighboring Mn ions in the planes ($J_1 > 0$) and for the exchange coupling between the next nearest neighboring Mn ions between these planes ($J_2 < 0$). J_{dij} is the exchange interaction constant for the doped case, which, due to the different ionic radii between the Sr and La ions, can be changed compared to the undoped case J_{ij}. $D_z > 0$ is the single-site anisotropy parameter of the easy-axis type.

A strong spin–phonon interaction in LMO is reported [9,11], which must be taken into account in order to obtain the correct results:

$$H^{sp-ph} = -\frac{1}{2}\sum_{i,j,k} F(i,j,k)Q_i S_j^z S_k^z - \frac{1}{4}\sum_{i,j,r,s} R(i,j,r,s)Q_i Q_j S_r^z S_s^z + h.c. \tag{2}$$

The normal coordinate Q_i can be expressed in terms of phonon creation a^+ and annihilation a operators $Q_i = (2\omega_{0i})^{-1/2}(a_i + a_i^+)$. ω_{0i} is the frequency of the lattice mode. F and R are the spin–phonon coupling constants in the first and second order, respectively. The spin–phonon interaction renormalizes the spin exchange interaction constant J to $J_{eff} = J + 2F^2/(\omega_o - MR)$. M is the magnetization.

H_{ph} contains the lattice vibrations, including anharmonic phonon–phonon interactions:

$$H_{ph} = \frac{1}{2!}\sum_i \omega_{0i} a_i^+ a_i + \frac{1}{3!}\sum_{i,j,r} B(i,j,r) Q_i Q_j Q_r + \frac{1}{4!}\sum_{i,j,r,s} A(i,j,r,s) Q_i Q_j Q_r Q_s. \tag{3}$$

A and B are three-phonon and four-phonon anharmonic interaction constants, respectively.

For the approximate calculation of the Green's functions, we use the method proposed by Tserkovnikov [25]. It goes beyond random phase approximation, taking into account all correlation functions. Moreover, this method allows us to calculate the imaginary part of the Green's function. We wish now to sketch it briefly. After the formal integration of the equation of motion for the retarding two-time Green's function

$$G_{ij}(t) = \langle\langle a_i(t); a_j^+ \rangle\rangle \tag{4}$$

one obtains

$$G_{ij}(t) = -i\theta(t)\langle [a_i; a_j^+] \rangle \exp(-i\omega_{ij}(t)t), \tag{5}$$

with $\theta(x) = 1$ for $x > 0$, $\theta(x) = 0$ for $x < 0$,

$$\omega_{ij}(t) = \omega_{ij} - \frac{i}{t}\int_0^t dt' t' \left(\frac{\langle [j_i(t); j_j^+(t')] \rangle}{\langle [a_i(t); a_j^+(t')] \rangle} - \frac{\langle [j_i(t); a_j^+(t')] \rangle \langle [a_i(t); j_j^+(t')] \rangle}{\langle [a_i(t); a_j^+(t')] \rangle^2} \right) \tag{6}$$

and $j_i(t) = \langle [a_i, H_{interaction}] \rangle$. The time-independent term

$$\omega_{ij} = \frac{\langle [[a_i, H]; a_j^+] \rangle}{\langle [a_i; a_j^+] \rangle} \tag{7}$$

is the excitation energy in the generalized Hartree–Fock approximation. The time-dependent term in Equation (7) includes damping effects.

We obtain from the full Hamiltonian the following expression for the phonon energy, which is renormalized through the spin–phonon interactions:

$$\omega_{ij}^2 = \omega_0^2 - 2\omega_0\Big(M_i M_j R_{ij}\delta_{ij} - \frac{1}{2N}\sum_r A_{ijr}^{ph}(2\bar{N}_r + 1) - B_{ij}^{ph}\langle Q_{ij}\rangle\delta_{ij} \Big), \tag{8}$$

with

$$\langle Q_{ij} \rangle = \frac{M_i M_j F_{ij}\delta_{ij} - \frac{1}{N'}\sum_r B_{ijr}^{ph}(2\bar{N}_r + 1)}{\omega_0 - M_i M_j R_{ij}\delta_{ij} + \frac{1}{N'}\sum_r A_{ijr}^{ph}(2\bar{N}_r + 1)}. \tag{9}$$

The correlation function of the phonons $\bar{N}_i = \langle a_i^+ a_i^- \rangle$ is obtained via the spectral theorem. M_i is the magnetization, which is calculated from the Green's function $\langle\langle S_i^+; S_j^- \rangle\rangle$. We obtain, for an arbitrary spin S value,

$$M = \langle S^z \rangle = \frac{1}{N}\sum_i \Big[(S + 0.5)\coth[(S + 0.5)\beta E_i] - 0.5\coth(0.5\beta E_i) \Big]. \tag{10}$$

E_i are the spin wave energies, $\beta = 1/k_B T$.

The phonon damping is also calculated taking into account the anharmonic spin–phonon and phonon–phonon interactions:

$$\gamma = \gamma_{sp-ph} + \gamma_{ph-ph}. \tag{11}$$

γ_{sp-ph} is the damping due to the spin–phonon interaction:

$$\begin{aligned}
\gamma_{sp-ph}(rs) &= \frac{2\pi}{N^2}\sum_{ij} F^2_{ijr}\langle S^z_i\rangle\langle S^z_j\rangle(\bar{n}_i - \bar{n}_j)\delta_{rs}\delta(E_i - E_j - \omega_r) \\
&+ \frac{2\pi}{N^3}\sum_{ijl} R^2_{ijlr}\Big(\langle S^z_i\rangle^2\langle S^z_j\rangle^2\delta_{rs} + [(\bar{n}_i - \bar{n}_j)(1+\bar{N}_l) + \bar{n}_i(1+\bar{n}_j)]\Big) \\
&* \delta(E_i - E_j - \omega_l + \omega_r).
\end{aligned} \tag{12}$$

γ_{ph-ph} is the damping part that arises from the phonon–phonon interaction:

$$\begin{aligned}
\gamma_{ph-ph}(rs) &= \frac{3\pi}{N^2}\sum_{ij} B^2_{ijr}(\bar{N}_i - \bar{N}_j)\delta_{rs}[\delta(-\omega_i - \omega_j + \omega_r) - \delta(\omega_i - \omega_j + \omega_r)] \\
&+ \frac{4\pi}{N^3}\sum_{ijl} A^2_{ijlr}\delta_{rs}[\bar{N}_i(1 + \bar{N}_j + \bar{N}_l) - \bar{N}_j\bar{N}_l] \\
&* \delta(\omega_i - \omega_j + \omega_l - \omega_r).
\end{aligned} \tag{13}$$

E_i and ω_i are the magnetic and phonon energies. The correlation functions $\bar{n}_i = \langle S^-_i S^+_i\rangle$ and $\bar{N}_i = \langle a^+_i a^-_i\rangle$ are obtained via the spectral theorem. It must be mentioned that in the damping γ at low temperatures, the terms due to the anharmonic spin–phonon interaction play an important role, whereas, at higher temperatures, above the phase transition temperature, only the anharmonic phonon–phonon terms remain.

In order to calculate the electrical conductivity σ, we need the Hubbard model:

$$H = \sum_{ij\sigma} t_{ij}c^+_{i\sigma}c_{j\sigma} + U\sum_i n_{i\uparrow}n_{i\downarrow}, \tag{14}$$

where t_{ij} is the hopping integral, U is the Coulomb repulsion, $n_{i\sigma} = c^+_{i\sigma}c_{i\sigma}$ and $c^+_{i\sigma}$ and $c_{i\sigma}$ are Fermi operators.

The electrical conductivity σ can be observed from the following equation:

$$\sigma^{\mu\nu}(\omega) - \frac{ie}{V}lim_{\delta\to 0^+}\sum_{ij\sigma} t_{ij}(R^\mu_i - R^\mu_j) \ll c^+_{i\sigma}c_{j\sigma}; P^\nu \gg_{\omega+i\delta}, \tag{15}$$

with $P^\nu = -e\sum i\sigma R^\nu_i n_{i\sigma}$; e is the electron charge and V is the volume.

We consider the Green's function $G_{ij\sigma}(\omega) = \ll c^+_{i\sigma}c_{j\sigma}; P^\nu \gg$, which can be written with the components

$$G_{ij\sigma}(\omega) = \sum_{\alpha\beta} G^{\alpha\beta}_{ij\sigma}(\omega) = \sum_{\alpha\beta} \ll c^+_{i\sigma}c_{j\sigma}n^\alpha_{i-\sigma}n^\beta_{j-\sigma}; P^\nu \gg. \tag{16}$$

For $G^{\alpha\beta}_{ij\sigma}(\omega)$, we obtain

$$G^{\alpha\beta}_{ij\sigma}(\omega) = \frac{e(R^\nu_j - R^\nu_i) < c^+_{i\sigma}c_{j\sigma}n^\alpha_{i-\sigma}n^\beta_{j-\sigma} >}{\omega + U(\delta_{\alpha+} - \delta_{\beta+})}. \tag{17}$$

After the calculation of the correlation function using the spectral theorem, from Equations (12) and (13), we can observe the electrical conductivity σ.

3. Numerical Results and Discussion

The numerical calculations are performed in the JAVA programming environment using simple iterative procedures and summation over nearest neighbors. They are performed using the following model parameters: $J_1 = 9.6$ K, $J_2 = -6.7$ K, $D_z = 1.92$ K [26], $S = 2$, $F = 23$ cm^{-1}, $R = -18$ cm^{-1}, $B = -2.54$ cm^{-1}, $A = 6.61$ cm^{-1}, $t = 1$ eV, $U = 2$ eV. The phonon–phonon interaction constants A and B are determined from the Raman spectra for temperatures above the Curie temperature T_C (where the terms with R and F vanish), whereas the spin–phonon interaction constants F and R—arise at very low temperatures—taking two values at two different temperatures from the Raman phonon energy and solving the system of two equations with two unknown parameters.

The exchange interaction constant J depends on the distance between the spins, i.e., on the lattice parameter, on the different strains, on the lattice symmetry and on the number of next neighbors. It is inversely proportional to the distance between two neighboring spins. The ionic radius of the doping ion Sr^{2+} is r(Sr^{2+}) = 1.44 Å, which is different compared to the host ion, r(La^{3+}) = 1.36 Å. This means that, in this case, where r(Sr) > r(La), there appears a tensile strain. For the exchange interaction constant in the doped case J_d, we use the relation $J_d < J$. An increase in the lattice parameters and the cell volume was observed for Sr-doped LMO by Zheng et al. [27]. The doping concentration is taken into account by the factor $(1-x)$ in Equation (1). For $x = 0$, without doping ions, J is the same as in the undoped compound and is maximal; then, with increasing x, J decreases, whereas J_d increases. For $x = 1$, where all host ions are substituted with doping ions, the first term in Equation (1) vanishes. Only the second term with J_d remains, which, for $x = 0$, is zero. This tensile strain, which reduces the exchange interaction constant, leads to a decrease in magnetization in Sr-doped LMO, in agreement with the experimental data of Wang et al. [28].

3.1. *Film Thickness Dependence of the Phonon Energy and Damping of the A_{1g} Mode 495 cm^{-1} in Pure and Sr-Doped LMO Thin Films*

Firstly, we investigate the phonon energy and damping for the A_{1g} mode, $\omega = 495$ cm^{-1}, in pure and Sr-doped free LMO thin films. The film thickness dependence of the phonon mode for a pure ($x = 0$) LMO thin film is calculated. It must be noted that the oxygen excess in LMO films is compensated by a mixed valence state of the manganese cation (Mn^{3+}/Mn^{4+}). The exchange interaction constants on the two free surfaces ($n = 1, N$) J_s are different from those in the bulk J; they can be larger or smaller. We have chosen the relation $J_s > J$, because Kharlamova et al. [29] have reported that, in LMO nanoparticles, the unit cell volume decreases. Then, through the spin–phonon interaction, we have $\mid R_s \mid > \mid R \mid$, $R < 0$. The phonon energy ω, which corresponds to the Raman peak position, increases with the decreasing film thickness d, and the distance between the layers is taken to be around 1 nm (see Figure 1, curve 1). This means that the Raman mode positions shift towards a higher frequency compared to the bulk single crystal reported for pure $LaMnO_3$ thin films [15]. The results show a remarkable difference in the phonon energies with the thickness. This behavior can be explained on the basis of the surface strain, which increases with decreasing thickness. As the thickness is increased, the phonon energies appear close to those of the single crystal.

Let us emphasize that the in-plane lattice parameters in LMO thin films are compressed, for example, by $LaAlO_3$ and $KTaO_3$ substrates [30]. This could be explained in our model by the exchange interaction constants between the film surface J_s and the substrate J_{ss}, $J_{ss} > J_s$; the phonon energy is enhanced in comparison to the case without a substrate (see Figure 1, curve 1s). For substrates that lead to a tensile in-plane strain (for example, MgO [30,31]), we must use the relation $J_{ss} < J_s$. The phonon energy is reduced compared to the free thin film. The effects of the substrate-induced strain on the phonon modes in Sr-doped LMO thin films [12–15] will be considered in a future paper.

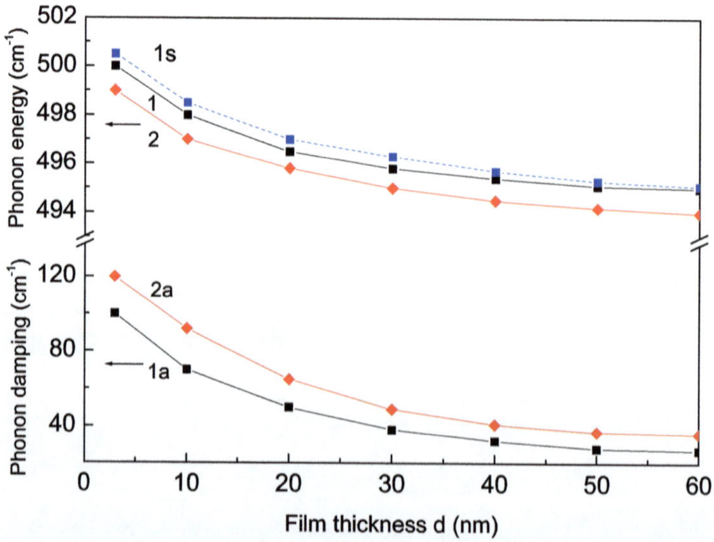

Figure 1. Film thickness dependence of the phonon energy of the A_{1g} mode ω = 495 cm^{-1} and their damping for T = 80 K in (1,1a) pure and (2,2a) Sr (x = 0.1) doped LMO thin film; (1s) LMO thin film on $LaAlO_3$ substrate.

The film thickness dependence of the phonon energy $\omega(N)$ for a Sr-doped LMO thin film, x = 0.1, is also observed using the model parameters $J_s > J$, $\mid R_s \mid > \mid R \mid$, $J_d < J$ and $\mid R_d \mid < \mid R \mid$, $R < 0$ (Figure 1, curve 2). The phonon energy ω increases again with the decreasing film thickness d, in agreement with the experimental data of Dore et al. [32]. The Sr doping could not change this behavior. However, it is observed that ω is smaller for the doped case (curve 2) compared to the undoped one (curve 1). This decreasing of ω for the Sr-doped LMO thin film is due to the different radii of the doped and host ions, due to the tensile strain, which appears after Sr doping in the LMO thin film. A similar decrease in ω with Sr is reported by many authors [10,11,20,21].

The phonon damping γ for pure and Sr-doped LMO thin films is also calculated. γ corresponds to the full width at half maximum (FWHM) of the Raman peaks. The results are shown in Figure 1. It can be seen that the phonon damping γ increases with the decreasing film thickness d. This means that the line widths of the Raman peaks also increase with the decreasing film thickness. Let us emphasize that γ is larger for the Sr-doped LMO thin film (curve 2a) than the undoped one (curve 1a). This is so because the doping contribution is additive to the surface one in the film, i.e.,

$$\gamma = \gamma_{bulk} + \gamma_{surface} + \gamma_{doping} + \dots \tag{18}$$

The broadening of the Raman peaks is reported after decreasing the film thickness in pure [15] and Sr-doped [32] LMO thin films. Let us emphasize that the distortion caused by the motions of oxygen atoms in Mn-O6 octahedra around the Mn ion are responsible for the Raman active vibrations. In doped manganites, the intensity and width of the broad bands are related to the amplitude of the dynamic fluctuations. The width of the instantaneous distribution of the Mn-O distances is the origin of the width of the Raman peaks. Therefore, as the Sr content increases, the activation energy decreases as well as the amplitude of the dynamic distortions. The width of the peaks corresponding to the $RMnO_3$-type spectrum increases as the Sr content rises. Moreover, from a structural point of view, the doping by substitution violates the translational invariance, which naturally causes an increase in the scattering of the phonon modes, i.e., the damping, of the FWHM.

3.2. *Temperature Dependence of the Phonon Energy and Damping of the A_{1g} Mode 495 cm^{-1} in Pure and Sr-Doped LMO Thin Films*

The temperature dependence of the phonon energy for the A_{1g} mode ω = 495 cm^{-1} in pure and Sr-doped LMO thin films is observed for Sr doping concentration $x = 0.1$ and using the following model parameters: $J_s > J$, $J_d < J$ and $| R_d | < | R |$, $R < 0$. The results are presented in Figure 2. The phonon energy decreases with an increasing temperature. There is a kink at T_N = 140 K in $\omega(T)$ for the pure LMO due to the spin–phonon interaction (curve 1). A similar kink was observed experimentally in pure LMO thin films by Dubey et al. [15] and Sr-doped bulk LMO by Bjoernsson et al. [21] and Podobedov et al. [10]. The kink in the Sr-doped LMO (which is already ferromagnetic) shifts to higher T_C values. It can be seen from Figure 2 that, for $x = 0.1$, the kink is at T_C = 200 K (curve 2). The increase in the Curie temperature of Sr-doped bulk LMO was reported also by Dimri et al. [33,34] and Ahmad et al. [35] using Raman and magnetic studies. The observed temperature behavior of Sr-doped LMO demonstrates the strong dependence of the phonons on both the temperature and the doping. The anomalies in the phonon energy are observed due to the effect of the ionic radius of the La/Sr site on both the doping and temperature. Thus, the results could be interpreted by considering the strong spin–phonon coupling in these compounds.

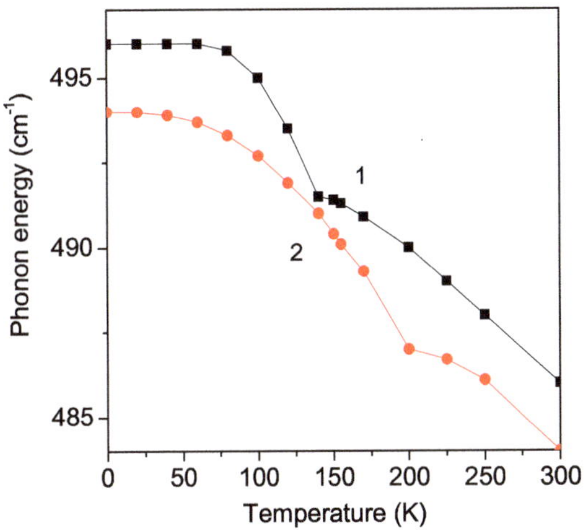

Figure 2. Temperature dependence of the phonon energy of the A_{1g} mode ω = 495 cm^{-1} in (1) pure and (2) Sr ($x = 0.1$) doped LMO thin film, d = 20 nm.

The phonon damping γ increases strongly with an increasing temperature T. This is supported by the experimentally observed broadening of the Raman peaks in these substances by Granado et al. [36]. There is an anomaly in the temperature dependence of the phonon damping γ of the A_{1g} mode ω = 495 cm^{-1} in the pure and Sr-doped LMO thin film around the phase transition temperatures: T = 140 K and T = 200 K for the undoped and Sr-doped LMO for $x = 0.1$ (see Figure 3). The anomalies in the phonon damping around the phase transition temperature T_N in the pure and Sr-doped LMO are due to the strong anharmonic spin–phonon interaction R. It must be noted that T_N is larger in the Sr-doped LMO than in the undoped one. The broadening of the Raman peaks in Sr-doped bulk LMO was reported by Podobedov et al. [10,37] and Choi et al. [11].

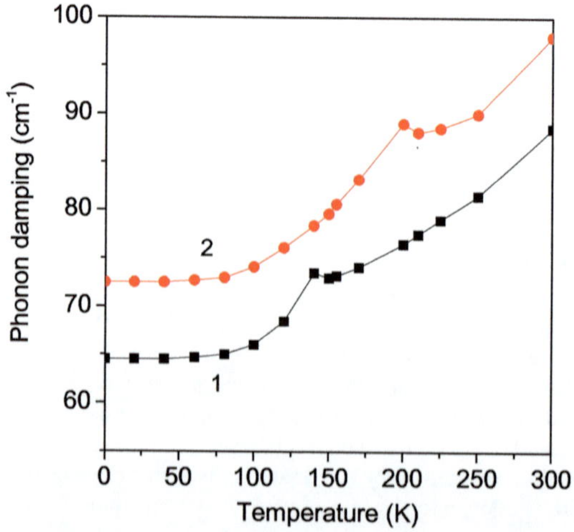

Figure 3. Temperature dependence of the phonon damping of the A_{1g} mode ω = 495 cm^{-1} in (1) pure and (2) Sr (x = 0.1) doped LMO thin film, d = 20 nm.

3.3. *Temperature Dependence of the Phonon Energy and Damping of the B_{2g} Mode 614 cm^{-1} in Pure and Sr-Doped LMO Thin Films*

Now, we will investigate the phonon energy ω of the B_{2g} mode with ω = 614 cm^{-1}. The experimental data of Dubey et al. [15] and Podobedov et al. [10] show an increase in ω with increasing temperature T. In order to obtain this hardening of ω, we choose a positive anharmonic spin–phonon constant $R > 0$ [38], $J_d < J$ and $R_d < R$. The results are demonstrated in Figure 4. There is again a kink at the phase transition temperatures: T = 140 K for the undoped case and T = 225 K for the Sr-doped LMO thin film for x = 0.1. Moreover, above the kinks, and above the phase transition temperatures, the phonon energies ω decrease slightly with an increasing temperature. This is due to the fact that, above T_C, only the anharmonic phonon–phonon interactions A and B remain. The anharmonic spin–phonon interaction with negative $R < 0$ and positive $R > 0$ causes the different behavior of ω—softening or hardening below the phase transition temperature.

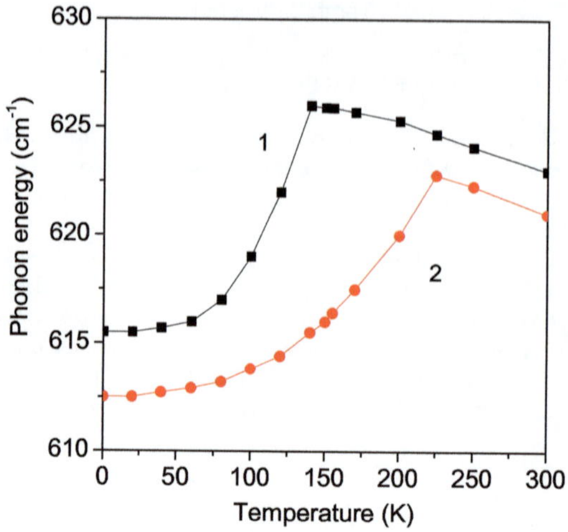

Figure 4. Temperature dependence of the phonon energy of the B_{2g} mode ω = 614 cm^{-1} in (1) pure and (2) Sr (x = 0.1) doped LMO thin film, d = 20 nm.

The phonon damping γ increases with temperature T (see Figure 5). There is again an anomaly around the phase transition temperatures: T = 140 K for the undoped case and T = 225 K for the Sr-doped LMO thin film for $x = 0.1$. It must be mentioned that γ is larger in the doped case compared to the undoped one for both cases—negative $R < 0$ and positive $R > 0$ (compare with Figure 4)—because the damping is proportional to R^2; see Equation (9). This behavior is due to the strong spin–phonon interaction R. An alternative means to observe the hardening of the phonon mode is the decrease in the Jahn–Teller distortion with increasing doping concentration x [13,36]. The reduction in the Jahn–Teller distortion in Sr-doped LMO has the same effect on the phonon energy as the change in the crystal symmetry due to the temperature-induced phase transition. This assumption is also confirmed by the polarized Raman spectra of doped crystals reported by Podobedov et al. [37]. Temperature-dependent Raman characterization has shown a metallic phase with a total reduction in the Jahn–Teller distortion in the rhombohedral $La_{0.67}Sr_{0.33}MnO_3$ nanoparticles [18].

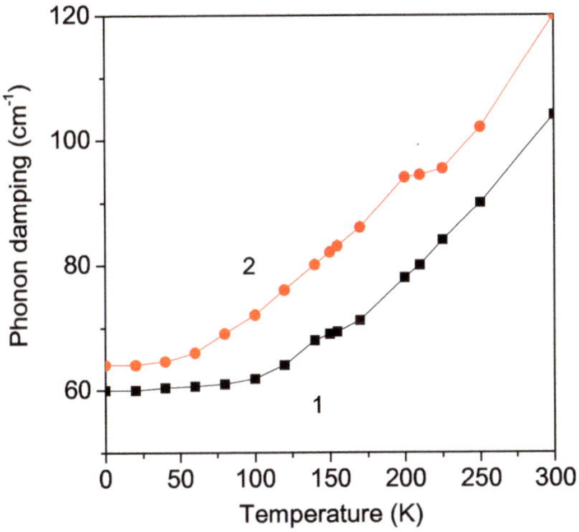

Figure 5. Temperature dependence of the phonon damping of the B_{2g} mode $\omega = 614$ cm^{-1} in (1) pure and (2) Sr ($x = 0.1$) doped LMO thin film, d = 20 nm.

3.4. Sr Doping Effect in LMO Thin Films on the Electrical Conductivity

Among perovskites with an ABO_3 structure, LMO is well known for solid oxide fuel cell (SOFC) applications. Strong electrical conductivity is necessary for the compound to serve as a cathode in SOFCs. Therefore, the doping of alkaline earth metals at the A-side enhances the electrical conductivity of the material [39]. Finally, from Equations (12)–(14), we have calculated the Sr concentration dependence of the electrical conductivity σ in an LMO thin film, d = 20 nm, T = 750 K. The substitution of La^{3+} with Sr^{2+} with a larger ionic radius favors both the crystal symmetry and the electrical conductivity of LMO. The results are presented in Figure 6. It can be seen that σ is enhanced when increasing the dopant concentration x; thus, it could be used for energy applications. This behavior is in qualitative agreement with the recently reported experimental results of Gupta et al. for Sr-doped LMO nanoparticles [4]. It must be noted that Gupta et al. [4] observed the highest σ increasing for $x = 0.15$. A similarly enhanced σ was obtained in Sr (on the La site) and Fe (on the Mn site) doped LMO by Shafi et al. [40] and transition metal-doped $LaMn_xM_{(1-x)}O_3$ (M = Fe, Co, Cr, Mn) [41]. Recently, He et al. [42] studied the electrical conductivity in Sr-doped $LaMO_3$ (M = Al, Ga, In, Er and Y) under varying oxygen partial pressures.

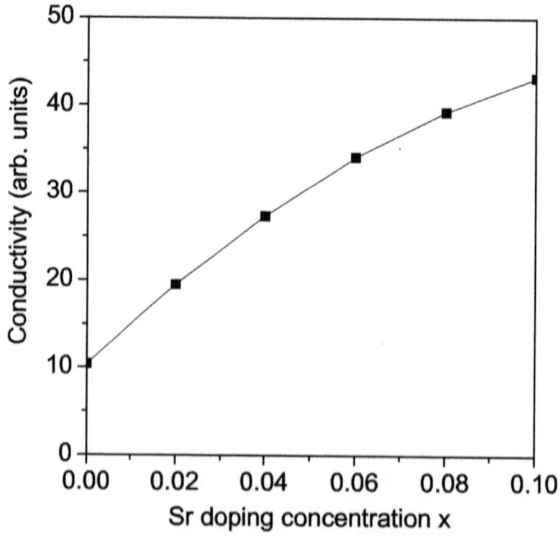

Figure 6. Sr doping concentration dependence of the electrical conductivity σ for LMO, d = 20 nm.

4. Conclusions

In conclusion, the film thickness, temperature, substrate and doping dependence of the phonon energy ω and damping γ of pure and Sr-doped LMO was investigated using a microscopic model and the Green's function theory. The kink near T_N = 140 K in pure LMO (due to strong spin–phonon interactions) shifts to higher T values with increasing Sr dopant. The softening and hardening of the ω = 495 cm^{-1} (A_{1g}) and ω = 614 cm^{-1} (B_{2g}) modes is explained by the different sign of the anharmonic spin–phonon interaction constant R, negative for the first case and positive for the second one. The damping increases with T for both cases, $R < 0$ and $R > 0$. In the temperature dependence of γ, there appears again an anomaly around the critical temperature. The phonon energy of the A_{1g} mode with ω = 295 cm^{-1} decreases, whereas the damping increases with an increasing Sr concentration. ω and γ increase strongly with the decreasing film thickness. The substrates can change also the phonon properties. The electrical conductivity is enhanced after the doping of LMO thin films with Sr, which could be used for energy storage applications.

Let us emphasize that the band gap is an important parameter in photocatalysts. Sr-doped LMO nanoparticles with low band gap energy of 2.2 eV could be used for energy applications [43]. The doping dependence of LMO with Sr and other doping ions will be investigated in a future paper.

Author Contributions: Conceptualization, J.M.W.; Methodology, A.T.A.; Software, I.N.A.; Formal analysis, A.T.A.; Investigation, A.T.A., I.N.A. and J.M.W.; Writing—original draft, J.M.W. All authors have read and agreed to the published version of the manuscript.

Funding: A.T.A. and I.N.A. acknowledge the financial support of the Bulgarian National Science Fund (contract number KP-06 PN68/17/BG-175467353-2022-04-0232).

Institutional Review Board Statement: Not applicable.

Informed Consent Statement: Not applicable.

Data Availability Statement: Data are contained within the article.

Conflicts of Interest: The authors declare no conflicts of interest.

References

1. Koriba, I.; Lagoun, B.; Guibadj, A.; Belhadj, S.; Ameur, A.; Cheriet, A. Structural, electronic, magnetic and mechanical properties of three $LaMnO_3$ phases: Theoretical investigations. *Comput. Condens. Matter* **2021**, *29*, e00592. [CrossRef]
2. Khomskii, D.I.; Sawatzky, G.A. Interplay between spin, charge and orbital degrees of freedom in magnetic oxides. *Solid State Commun.* **1997**, *102*, 87. [CrossRef]
3. Hess, F.; Yildiz, B. Precipitation of dopants on acceptor-doped $LaMnO_3$ revealed by defect chemistry from first principles. *J. Chem. Phys.* **2021**, *154*, 064702. [CrossRef] [PubMed]
4. Gupta, K.; Thakur, O.P.; Kumar, M. Effect of A-Site Substitution on $LaMnO_3$ Perovskite via Sr Ions for Energy Applications. *J. Electron. Mater.* **2023**, *52*, 4279–4288. [CrossRef]
5. Marton, Z.; Seo, S.S.A.; Egami, T.; Lee, H.N. Growth control of stoichiometry in $LaMnO_3$ epitaxial thin films by pulsed laser deposition. *J. Cryst. Growth* **2010**, *312*, 2923. [CrossRef]
6. Wesselinowa, J.M. Size and anisotropy effects on magnetic properties of antiferromagnetic nanoparticles. *J. Magn. Magn. Mater.* **2010**, *322*, 234. [CrossRef]
7. Jonker, G.; van Santen, J. Ferromagnetic compounds of manganese with perovskite structure. *Physica* **1950**, *16*, 337. [CrossRef]
8. Kovaleva, N.N.; Kusmartseva, O.E.; Kugel, K.I.; Maksimov, A.A.; Nuzhnyy, D.; Balbashov, A.M.; Demikhov, E.I.; Dejneka, A.; Trepakov, V.A.; Kusmartsev, F.V.; et al. Anomalous multi-order Raman scattering in $LaMnO_3$: A signature of quantum lattice effects in a Jahn–Teller crystal. *J. Phys. Condens. Matter* **2013**, *25*, 155602. [CrossRef] [PubMed]
9. Sidorov, T.A. Identification of complex anions in La1-xAxMnO$_3$ manganites (A = Ca, Sr) from neutron diffraction data and refinement of their structures on the basis of raman spectroscopy data. *Russian J. Inorg. Chem.* **2013**, *58*, 706. [CrossRef]
10. Podobedov, V.B.; Weber, A.; Romero, D.B.; Rice, J.P.; Drew, H.D. Raman scattering in La1-xSrxMnO$_3$ single crystals (x = 0, 0.1, 0.2, 0.3). *Solid State Commun.* **1998**, *105*, 589. [CrossRef]
11. Choi, K.-Y.; Lemmens, P.; Sahaoui, T.; Güntherodt, G.; Pashkevich, Y.G.; Gnezdilov, V.P.; Reutler, P.; Pinsard-Gaudart, L.; Buechner, B.; Revcolevschi, A. Existence of orbital polarons in ferromagnetic insulating La1-xSrxMnO$_3$(0.11 $< x <$ 0.14) revealed by giant phonon softening. *Phys. Rev. B* **2005**, *71*, 174402. [CrossRef]
12. Aruta, C.; Angeloni, M.; Balestrino, G.; Boggio, N.G.; Medaglia, P.G.; Tebano, A.; Davidson, B.; Baldini, M.; Di Castro, D.; Postorino, P.; et al. Preparation and characterization of $LaMnO_3$ thin films grown by pulsed laser deposition. *J. Appl. Phys.* **2006**, *100*, 023910. [CrossRef]
13. Safarina, G.A.; Kim, Y.-J.; Park, H.-S.; Yang, C.-H. Raman spectroscopy of the Jahn-Teller phonons in a magnetic $LaMnO_3$ thin film grown on $KTaO_3$. *J. Appl. Phys.* **2022**, *131*, 025302. [CrossRef]
14. Chaturvedi, A.; Sathe, V. Thickness dependent Raman study of epitaxial $LaMnO_3$ thin films. *Thin Solid Films* **2013**, *548*, 75. [CrossRef]
15. Dubey, A.; Sathe, V.G. The effect of magnetic order and thickness in the Raman spectra of oriented thin films of $LaMnO_3$. *J. Phys. Condens. Matter* **2007**, *19*, 346232. [CrossRef]
16. Hernandez, E.; Sagredo, V.; Delgado, G.E. Synthesis and magnetic characterization of $LaMnO_3$ nanoparticles. *Rev. Mex. Fis.* **2015**, *61*, 166. Available online: https://api.semanticscholar.org/CorpusID:56081270 (accessed on 22 April 2024).
17. Rodriguez-Lamas, R.; Pla, D.; Chaix-Pluchery, O.; Meunier, B.; Wilhelm, F.; Rogalev, A.; Rapenne, L.; Mescot, X.; Rafhay, O.; Roussel, H.; et al. Integration of $LaMnO_3$ films on platinized silicon substrates for resistive switching applications by PI-MOCVD. *Beilstein J. Nanotechn.* **2019**, *10*, 389. [CrossRef]
18. Saleem, M.; Varshney, D. Structural, thermal, and transport properties of La0.67Sr0.33MnO3 nanoparticles synthesized via the sol-gel auto-combustion technique. *RSC Adv.* **2018**, *8*, 1600. [CrossRef]
19. Helton, J.S.; Zhao, Y.; Shulyatev, D.A.; Lynn, J.W. Damping and softening of transverse acoustic phonons in colossal magnetoresistive $La_{0.7}Ca_{0.3}MnO_3$ and $La_{0.7}Sr_{0.3}MnO_3$. *Phys. Rev. B* **2019**, *99*, 024407. [CrossRef]
20. Mayr, F.; Hartinger, C.; Loidl, A. Structural aspects of the phonon spectra of La1-xSrxMnO$_3$. *Phys. Rev. B* **2005**, *72*, 024425. [CrossRef]
21. Bjornsson, P.; Ruebhausen, M.; Backstrom, J.; Kall, M.; Eriksson, S.; Eriksen, J.; Borjesson, L. Lattice and charge excitations in La1-xSrxMnO$_3$. *Phys. Rev. B* **2000**, *61*, 1193. [CrossRef]
22. Wdowik, U.D.; Koza, M.M.; Chatterji, T. Phonons in lanthanum manganite: Inelastic neutron scattering and density functional theory studies. *Phys. Rev. B* **2012**, *86*, 174305. [CrossRef]
23. Talati, M.; Jha, P.K. Structure dependent phonon properties of $LaMnO_3$. *Comput. Mater. Sci.* **2006**, *37*, 64. [CrossRef]
24. Rini, E.G.; Rao, M.N.; Chaplot, S.L.; Gaur, N.K.; Singh, R.K. Phonon dynamics of lanthanum manganite $LaMnO_3$ using an interatomic shell model potential. *Phys. Rev. B* **2007**, *75*, 214301. [CrossRef]
25. Tserkovnikov, Y.A. Decoupling of chains of equations for two-time Green's functions. *Theor. Math. Phys.* **1971**, *7*, 511. [CrossRef]
26. Naji, S.; Benyoussef, A.; El Kenz, A.; Ez-Zahraouy, H.; Loulidi, M. Monte Carlo study of phase transitions and magnetic properties of $LaMnO_3$. *Phys. Stat. Mech. Its Appl.* **2011**, *13*, 38.
27. Zheng, F.; Pederson, L.R. Phase Behavior of Lanthanum Strontium Manganites. *J. Electrochem. Soc.* **1999**, *146*, 2810. [CrossRef]
28. Wang, T.; Wu, H.-Y.; Sun, Y.-B.; Xing, R.; Xv, B.; Zhao, J.-J. Physical Properties of Sr-Doped Double Perovskite La_2NiMnO_6. *J. Supercond. Nov. Magn.* **2020**, *33*, 727. [CrossRef]
29. Kharlamova, M.V.; Arulraj, A. Phase transition in nanostructured $LaMnO_3$. *JETP Lett.* **2009**, *89*, 301–305. [CrossRef]

30. Vukmirovic, J.; Joksovic, S.; Piper, D.; Nesterovic, A.; Novakovic, M.; Rakic, S.; Milanovic, M.; Srdic, V.V. Epitaxial growth of $LaMnO_3$ thin films on different single crystal substrates by polymer assisted deposition. *Ceram. Int.* **2023**, *49*, 2366–2372. [CrossRef]
31. Daoudi, K.; El-Helali, S.; Othmen, Z.; Suleiman, B.M.; Tsuchiya, T. Microstructure and electrical transport mechanisms of the Ca-doped $LaMnO_3$ films grown on MgO substrate. *J. Mater.* **2020**, *6*, 17–23. [CrossRef]
32. Dore, P.; Postorino, P.; Sacchetti, A.; Baldini, M.; Giambelluca, R.; Angeloni, M.; Balestrino, G. Raman measurements on thin films of the $La_{0.7}Sr_{0.3}MnO_3$ manganite: A probe of substrate-induced effects. *Eur. Phys. J. B* **2005**, *48*, 255. [CrossRef]
33. Dimri, M.C.; Khanduri, H.; Mere, A.; Stern, R. Studies of doped $LaMnO_3$ samples prepared by citrate combustion process. *AIP Conf. Proc.* **2018**, *1942*, 130015. [CrossRef]
34. Dimri, M.C.; Khanduri, H.; Stern, R. Effects of aliovalent dopants in $LaMnO_3$: Magnetic, structural and transport properties. *J. Magn. Magn. Mater.* **2021**, *536*, 168111. [CrossRef]
35. Ahmad, T.; Ramanujachary, K.V.; Lofland, S.E.; Ganguli, A.K. Reverse micellar synthesis and properties of nanocrystalline GMR materials ($LaMnO_3$, $La_{0.67}Sr_{0.33}MnO_3$ and $La_{0.67}Ca_{0.33}MnO_3$): Ramifications of size considerations. *J. Chem. Sci.* **2006**, *118*, 513–518. [CrossRef]
36. Granado, E.; Sanjurjo, J.A.; Rettori, C.; Neumeier, J.J.; Oseroff, S.B. Order-disorder in the Jahn-Teller transition of $LaMnO_3$: A Raman scattering study. *Phys. Rev. B* **2000**, *62*, 11304. [CrossRef]
37. Podobedov, V.B.; Weber, A.; Romero, D.L.; Rice, J.P.; Drew, H.D. Effect of structural and magnetic transitions in La1-xMxMnO_3 (M = Sr, Ca) single crystals in Raman scattering. *Phys. Rev. B* **1998**, *58*, 43. [CrossRef]
38. Wesselinowa, J.M.; Apostolov, A.T. Anharmonic effects in ferromagnetic semiconductors. *J. Phys. Condens. Matter* **1996**, *8*, 473. [CrossRef]
39. Paul, S.; Anand, V.; Noorjahan; Duggal, V.; Manokamna. Preparation and performance of a Sr^{2+} doped $LaMnO_3$ cathode for low temperature solid oxide fuel cells. *Mater. Today Proc.* **2023**. [CrossRef]
40. Shafi, P.M.; Mohapatra, D.; Reddy, V.P.; Dhakal, G.; Kumar, D.R.; Tuma, D.; Brousse, T.; Shim, J.-J. Sr- and Fe-substituted $LaMnO_3$ Perovskite: Fundamental insight and possible use in asymmetric hybrid supercapacitor. *Energy Storage Mater.* **2022**, *45*, 119–129. [CrossRef]
41. Okuyucu, H.; Cinici, H.; Konak, T. Coating of nano-sized ionically conductive Sr and Ca doped $LaMnO_3$ films by sol–gel route. *Ceram. Int.* **2013**, *39*, 903. [CrossRef]
42. He, C.; Wu, J.; Lee, Y. Correlation between conductivity and structural parameters in Sr-doped $LaMO_3$ (M = Al, Ga, In, Er, and Y) for solid oxide membranes. *Solid State Ionics* **2023**, *399*, 116315. [CrossRef]
43. Tran, T.H.; Phi, T.H.; Nguyen, H.N.; Pham, N.H.; Nguyen, C.V.; Ho, K.H.; Doan, Q.K.; Le, V.Q.; Nguyen, T.T.; Nguyen, V.T. Sr doped $LaMnO_3$ nanoparticles prepared by microwave combustion method: A recyclable visible light photocatalyst. *Results Phys.* **2020**, *19*, 103417. [CrossRef]

materials

Article

Multifunctional Superamphiphobic Coating Based on Fluorinated TiO_2 toward Effective Anti-Corrosion

Xiao Huang [1], Xinghua Gao [1], Xin Wang [2], Hongfei Shang [3] and Shujun Zhou [1,*]

[1] School of Mechanical and Electrical Engineering, China University of Mining and Technology (Beijing), Beijing 100083, China; hx@cumtb.edu.cn (X.H.); 17810286684@sina.cn (X.G.)

[2] Surface Engineering Institution, AECC Beijing Institute of Aeronautical Materials, Beijing 100095, China; rasheed990918@163.com

[3] State Key Laboratory of Tribology, Tsinghua University, Beijing 100084, China; shanghongfei@tsinghua.edu.cn

* Correspondence: zhoushujun@cumtb.edu.cn

Abstract: The application of superamphiphobic coatings improves the surface's ability to repel fluids, thereby greatly enhancing its various functions, including anti-fouling, anti-corrosion, anti-icing, anti-bacterial, and self-cleaning properties. This maximizes the material's potential for industrial applications. This work utilized the agglomeration phenomenon exhibited by nano-spherical titanium dioxide (TiO_2) particles to fabricate 1H,1H,2H,2H-perfluorodecyltriethoxysilane (PFDTES) modified TiO_2 (TiO_2@fluoroPOS) fillers with low surface energy. This was achieved through the in-situ formation of protective armor on the surface of the agglomerates using the sol-gel method and fluorination modification. Polyvinylidene fluoride-tetrafluoropropylene (PVDF-HFP) and TiO_2@fluoroPOS fillers were combined using a spraying technique to prepare P/TiO_2@fluoroPOS coatings with superamphiphobicity. Relying on the abundance of papillae, micropores, and other tiny spaces on the surface, the coating can capture a stable air film and reject a variety of liquids. When the coatings were immersed in solutions of 2 mol/L HCl, NaCl, and NaOH for a duration of 12 h, they retained their exceptional superamphiphobic properties. Owing to the combined influence of the armor structure and the organic binder, the coating exhibited good liquid repellency during water jetting and sandpaper abrasion tests. Furthermore, the coating has shown exceptional efficacy in terms of its ability to be anti-icing, anti-waxing, and self-cleaning.

Keywords: superamphiphobic; armor structure; spraying method; robust; multifunctional

Citation: Huang, X.; Gao, X.; Wang, X.; Shang, H.; Zhou, S. Multifunctional Superamphiphobic Coating Based on Fluorinated TiO_2 toward Effective Anti-Corrosion. *Materials* **2024**, *17*, 2203. https://doi.org/10.3390/ma17102203

Academic Editors: Sebastiano Vasi and Ulderico Wanderlingh

Received: 2 April 2024
Revised: 25 April 2024
Accepted: 6 May 2024
Published: 8 May 2024

1. Introduction

By improving a surface's ability to repel fluids, special anti-wetting coatings can maximize the material's potential for industrial applications by improving the surface's anti-fouling [1–3], anti-corrosion [3–6], anti-icing [6–10], anti-bacterial [7], and many other properties [11]. Researchers have used low surface energy materials to construct micro-nano roughness structures in order to prepare superhydrophobic coatings, which are widely used in antifouling, anti-corrosion, anti-icing, and self-cleaning applications [12–17]. These coatings are inspired by the unique wettability of plants and animals found in nature, such as lotus leaves [18], water striders' legs [19], rose petals [15], etc. The application of superhydrophobic coatings in oil-contaminated environments is restricted since these coatings, even with basic micro- and nano-rough structures, are frequently unable to repel oil and other liquids with a surface energy lower than water [20,21]. In order to expand the coating's wettability resistance from water to oil, the concept of superamphiphobic surfaces has emerged, i.e., the contact angle of both water and oil is greater than 150°, and the sliding angle of both water and oil is less than 10° [22–24]. and the generally accepted prerequisites for the development of superamphiphobic surfaces—that is, materials with extremely low surface energy and specifically designed surface structures [20], like kangaroo, overhanging [13], and re-entrant structures [25,26]—have been established. In order to create re-entrant structures on the surface of silicon wafers, Kang et al. [27] used lithography to produce mushroom-shaped micropillar arrays with great structural fidelity. It was discovered that these structures were super-liquid repellent to low surface energy liquids like white mineral oil. In order to create meter-like porous structured superamphiphobic coatings, Ganesh et al. [28] used PVA-TiO_2 nano-fibers on glass substrates and the electrostatic spinning method. The prepared surfaces' contact angles with ethylene glycol and water were 166° and 152.6°, respectively. Strict and exact control over the surface structure is necessary to construct such particularly formed superamphiphobic surfaces with high structural fidelity. This calls for the employment of labor-intensive and sophisticated processes like electrostatic spinning, photolithography, and reactive ion etching. These techniques have limited applicability, are frequently costly, and are dependent on the substrate material for the manufacture of superamphiphobic surfaces.

The most promising solution to this problem now available is the spraying method, which is easy to use, quick, affordable, and suitable for a wide range of large-area surfaces [29–32]. There is a dearth of literature on superamphiphobic coatings because the spray method builds up nano-particles on the coating's surface, and it is challenging to control the coating's morphology to meet the requirements of superamphiphobic microstructures [28,33,34]. In the literature that has been reported thus far, the coatings sprayed with nano-particle vegetation are primarily superhydrophobic coatings. This is because the requirements of superhydrophobic coatings on the surface morphology are relatively easy to meet. Wang [35] claims that, compared to nano-spherical particles with rough surfaces, the coating made by the spraying method using smooth nano-spheres has much less liquid repellency. Cassie's principle [21] states that the solid-liquid contact area and contact angle decrease with increasing filler particle surface roughness. Consequently, employing the spraying method to create superamphiphobic coating microstructures might be aided by creating superamphiphobic filler particles with rough surfaces.

Prior to coating application, the coating must have a suitable balance of mechanical stability and liquid repellency [17,36]. In order to make the coatings adaptable to more demanding environments, the researchers tried to introduce different kinds of binders to enhance the mechanical durability of the coatings [31,37]. The superamphiphobic surfaces formed only by the agglomeration and accumulation of nano-superamphiphobic fillers are almost always fragile and have poor mechanical durability [38]. Natural wear and tear or even a slight touch may destroy their texture structure and lead to the loss of their superamphiphobicity [17,20,29]. To create a multilayer superamphiphobic covering, Zhu [30] used transparent silicone resin as a binder and sprayed micro-nano-particles on top of the silicone resin. Zhou [19] was able to create a coating that could sustain superamphiphobicity under extreme stretching conditions by using PDMS as a binder and spraying silicon nanofilaments onto a substrate that had already been stretched. The surface structure exhibits poor stability and robustness as the addition of binder increases the bonding force between the coating and the substrate. However, the nano-particle stacking causes a small inter-particle gap that restricts the entry of binder, and the attraction between the agglomerated nano-particles is weak, resulting in a poor inter-particle bonding force [39]. The in-situ growth of protective shells on the surface of their agglomerates can preserve the rough structure of the nano-particle-stacked structure and increase the robustness of the structure by utilizing the agglomeration effect of the nano-particles. This increases the stability of the entire coating to capture the air film.

In this work, the thermoplastic polymer polyvinylidene fluoride-hexafluoropropylene (PVDF-HFP), which has good binding strength and corrosion resistance, was chosen as the binder, and nano-spherical titanium dioxide particles were chosen as the precursor for the synthesis of 1H,1H,2H,2H-perfluorodecyltriethoxysilane (PFDTES) modified TiO_2 (TiO_2@fluoroPOS) functional fillers. Titanium dioxide rough armors were grown in situ on the surface of titanium dioxide agglomerates using the sol-gel method, utilizing the agglomeration effect of the nano-particles. Then, using 1H,1H,2H,2H-perfluorodecyltriethoxysilane (PFDTES), the TiO_2 rough particles were fluorinated to produce TiO_2@fluoroPOS fillers with a high fluorine content on the surface. This process produced functional fillers with low surface energy and high roughness. After mixing and spraying the filler and binder, a multifunctional P/TiO_2@fluoroPOS superamphiphobic coating with outstanding stability was created. In terms of corrosion, self-cleaning, antifouling, anti-icing, and anti-waxing properties, the coating performed admirably.

2. Materials and Methods

2.1. Materials

Titanium dioxide (TiO_2, spherical particles, particle size 20~30 nm) and polyvinylidene fluoride-hexafluoropropylene (PVDF-HFP, 99%) were supplied by Aladdin, Wallingford, CT, USA. Tetrabutyl titanate (TBT, AR, 98%), acetic acid (CH_3COOH, 99.9%), 1H,1H,2H,2H-perfluorodecyltriethoxysilane (PFDTES, 97%), hydrochloric acid (HCl, AR, 36~38%), sodium hydroxide (NaOH, ACS, 97%), sodium chloride (NaCl, AR), and n-hexadecane (RG, 98%) were all supplied by Adamas, Emeryville, CA, USA. Anhydrous ethanol (EtOH, AR, 99.5%) and mineral oil (RG, 99%) were supplied by Greagent, Ringoes, NJ, USA. Soybean oil was supplied by Luhua, Yantai, China.

2.2. Preparation of TiO_2@fluoroPOS Fillers

The TiO_2@fluoroPOS superamphiphobic fillers were prepared by the sol-gel method, and the preparation process is shown in Figure 1a. Firstly, 2 g of TiO_2 powder was dispersed in a mixed solution of 200 mL of anhydrous ethanol, 20 mL of deionized water, and 12 mL of acetic acid, then 6 mL of TBT was slowly added dropwise to the solution and stirred vigorously for 6 h to ensure complete hydrolysis. Subsequently, 1.6 mL of PFDTES was added dropwise to the solution and reacted for 12 h. Finally, the TiO_2@fluoroPOS fillers were obtained after washing with anhydrous ethanol to a PH > 6.5 and drying at 70 °C for 12 h.

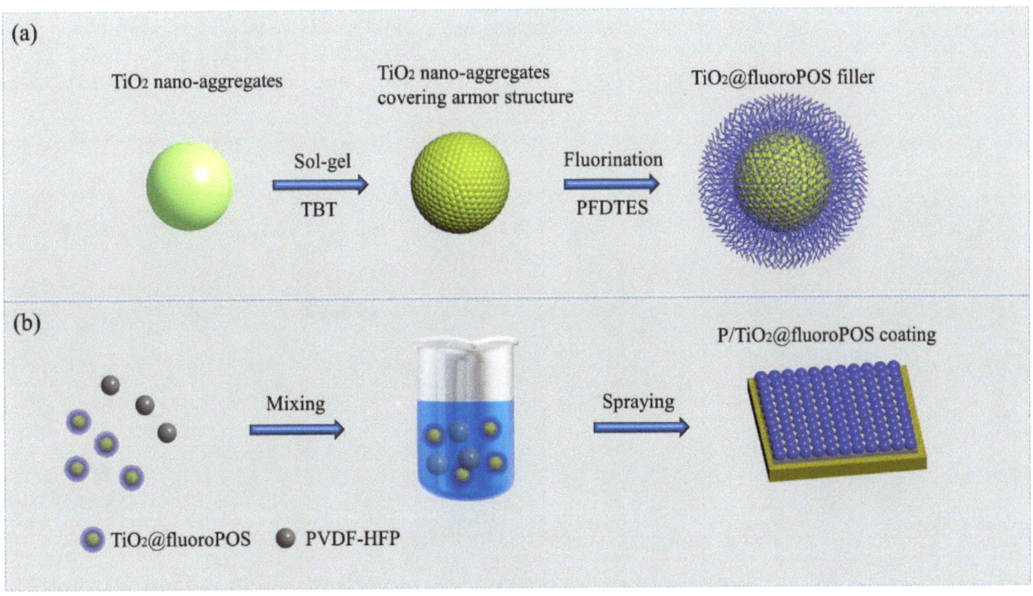

Figure 1. (**a**) The schematic illustration of the fabrication of filler TiO_2@fluoroPOS. (**b**) The schematic illustration of the preparation of the superamphiphobic coating P/TiO_2@fluoroPOS.

2.3. Preparation of P/TiO_2@fluoroPOS Coatings

The coatings were prepared by spraying with a mixture of filler and binder, and the preparation process is shown in Figure 1b. The Al substrate was polished using 1000 mesh sandpaper to remove the oxide layer and then cleaned with deionized water and anhydrous ethanol. Next, 2 g of TiO_2@fluoroPOS filler and 1.5 g of PVDF-HFP binder were dispersed in 20 mL of anhydrous ethanol. The dispersions were sprayed on the Al substrate (spraying area 24 cm^2) using a spray gun at a distance of about 20 cm from the Al substrate and a pressure of 0.3 MPa. The filler surface concentration ranged from 41.7 to 62.5 g/m^2. Finally, the superamphiphobic P/TiO_2@fluoroPOS coating was obtained after curing at 220 °C for 60 min.

2.4. Characterization

The contact angles (CAs) of water and oil were measured by a contact angle measuring system (ZJ-6900, ZJ, Shenzhen, China). The sliding angles (SAs) of water and oil were measured by angular tilting platforms (GFWG60-60, MISUMI, Shanghai, China). The measurement results were the average value of the CAs and SAs at five positions with 10 μL water droplets and oil droplets. The crystal structures of the particles were determined by X-ray diffraction (XRD, X'Pert PRO MPD, Nalytical, Alemlo, The Netherlands). The functional groups of the original TiO_2 and TiO_2@fluoroPOS were analyzed by Fourier transform infrared spectroscopy (FT-IR, Nicolet iS 10, Green Bay, WI, USA). The surface chemistry of TiO_2@fluoroPOS fillers was analyzed by X-ray photoelectron spectroscopy (XPS, Thermo Scientific K-Alpha, Waltham, MA, USA). Binding energies were calibrated with reference to the C1s peak at 284.8 eV. The surface morphology of the fillers and coatings was observed by scanning electron microscopy (SEM, Hitachi S3400N, Hitachi, Tokyo, Japan).

3. Results and Discussion

3.1. Analyses of TiO_2@fluoroPOS Filler

Figure 2 depicts the TiO_2@fluoroPOS filler's chemical reaction pathway. In an acidic environment, tetrabutyl titanate (TBT) readily hydrolyzes to form hydroxyl groups. Following a sequence of hydrolysis condensation reactions, TiO_2 armors form on the surface of TiO_2 agglomerates. The fluorine-containing groups of PFDTES are then grafted onto the surface of titanium dioxide via the condensation reaction, yielding a low-surface-energy functional filler.

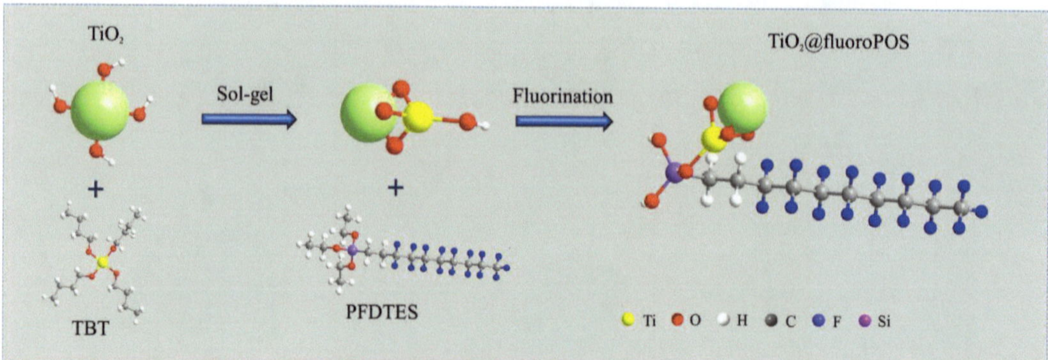

Figure 2. The reaction schematic for the fabrication of filler TiO_2@fluoroPOS.

Figure 3a displays the original TiO_2 and TiO_2@fluoroPOS XRD test results. Characteristic peaks of the original TiO_2 are located at 2θ = 25.36°, 37.83°, 8.09°, 53.90°, 55.0°, 62.74°, 68.81°, 70.38°, and 75.12°, respectively; they correspond to the crystal planes (101), (004), (200), (105), (211), (204), (116), (220), and (215). The reacted TiO_2@fluoroPOS characteristic peaks do not exhibit any discernible changes from the original TiO_2, and they align with the standard peak base of anatase TiO_2. This suggests that the TiO_2 armors grown on the surface of the TiO_2 agglomerates possess the same structure as the original TiO_2.

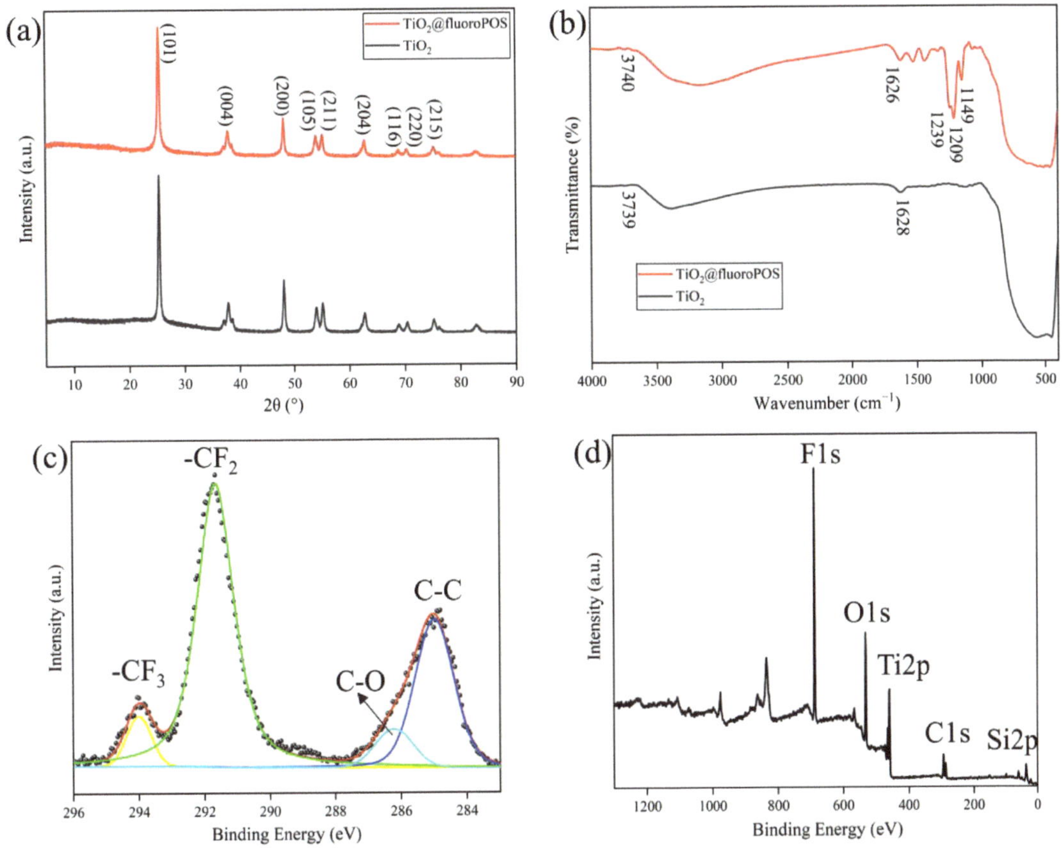

Figure 3. (**a**) XRD patterns of original TiO_2 and TiO_2@fluoroPOS; (**b**) FT-IR spectra of original TiO_2 and TiO_2@fluoroPOS; (**c**) XPS survey spectra of TiO_2@fluoroPOS; (**d**) high-resolution C 1s spectra of TiO_2@fluoroPOS.

Figure 3b displays the results of the Fourier transform infrared spectroscopy (FT-IR) measurements of the original TiO_2 and TiO_2@fluoroPOS. The primary cause of TiO_2's absorption peak at 3739 cm^{-1} is water from crystallization, while the vibration of immobilized hydroxyl groups on the surface of TiO_2 is responsible for the absorption peak at 1628 cm^{-1}. TiO_2@fluoroPOS's absorption peaks at 3740 cm^{-1} and 1626 cm^{-1} indicate the existence of hydroxyl groups and water crystallization. The three new absorption peaks at 1239 cm^{-1}, 1209 cm^{-1}, and 1149 cm^{-1} are due to the stretching vibrations of the -CF_3 and -CF_2 groups of PFDTES. This suggests that the groups containing fluorine have effectively branched onto the TiO_2 armor surface.

As seen in Figure 3c,d, XPS was used to confirm the chemical composition of TiO_2@ fluoroPOS. The presence of many fluorine elements on the surface of the TiO_2@fluoroPOS filler was indicated by the detection of a high peak of F1s at 689.05 eV. Furthermore, the -CF_3, -CF_2, C-O, and C-C groups of PFDTES were detected as the peaks at 294.0 eV, 291.7 eV, 286.2 eV, and 284.8 eV in the C1s high-energy spectra. In order to create low surface energy, which is essential for building the superamphiphobic coating, the results show that several low surface energy groups have been effectively branched onto the filler's surface.

3.2. Surface Morphology of TiO_2@fluoroPOS Fillers and P/TiO_2@fluoroPOS Coatings

Figure 4a displays the microstructures of the original TiO_2. The surface of the untreated TiO_2 is smooth and spherical, with individual particles having a diameter of roughly 20~30 nm. Van der Waals forces cause the TiO_2 nano-particles to aggregate into agglomerates, which have a rough, uneven surface morphology. Despite this, van der Waals forces do not strongly bond the particles together, and there are clear interstitial gaps between the particles. TiO_2@fluoroPOS's microstructure can be observed in Figure 4b,c. Agglomerated particles with armor structure are roughly 1~2 μm in diameter. The surface of the particles retains the original rough structure of the agglomerates, and the granular protrusions are tightly bound to encapsulate the agglomerates. During the sol-gel process, the titanium dioxide armored structure grows in situ along the agglomerate surface, encapsulating the agglomerates into a micrometer particle with a surface nano-roughness structure that is critical to the robustness and construction of the superamphiphobic coatings.

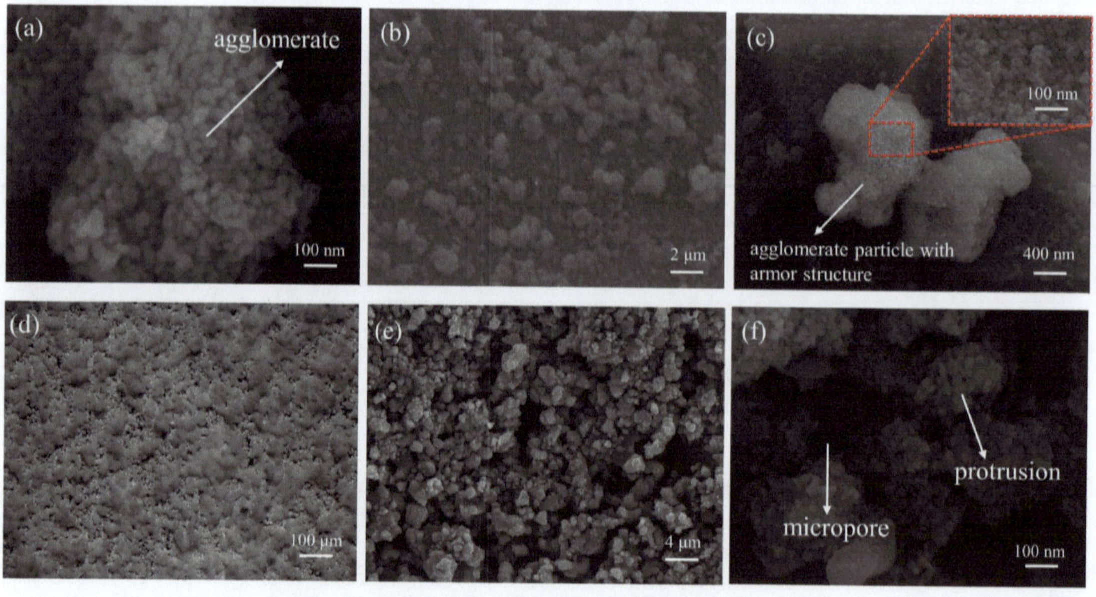

Figure 4. The SEM images of (**a**) TiO_2 and (**b**,**c**) TiO_2@fluoroPOS fillers. (**d**) Optical microscope photograph of P/TiO_2@fluoroPOS coatings. (**e**,**f**) The SEM images of P/TiO_2@fluoroPOS coatings.

The P/TiO_2@fluoroPOS coating's morphology is displayed in Figure 4d–f. The optical microscope image exhibits hilly bumps distributed on the surface of the coating. A solid and unique micro-nano roughness structure is formed by the fillers with a micro-nano roughness structure layered interleaved on the coating's surface, creating a lot of protrusions and micropores. The fillers are adhered to and supported by the polymer binder. The coating is in the Cassie–Baxter state, which facilitates the suspension of droplets on top of the rough surface and results in superamphiphobicity. It does this by relying on the numerous papillae, micropores, and other small gaps working together to capture a stable air film.

3.3. Wettability of P/TiO_2@fluoroPOS Coatings

The wettability of the Al substrate, PVDF-HFP coating, and P/TiO_2@fluoroPOS coating was assessed by measuring the contact angles of water and mineral oil. Figure 5a demonstrates that the naked Al surface is both hydrophilic and amphiphilic, but the PVDF-HFP surface is hydrophobic and lipophilic. The contact angles of the P/TiO_2@fluoroPOS coating for water and mineral oil were 160.8° and 157.4°, respectively. This demonstrates the coating's exceptional superamphiphobicity.

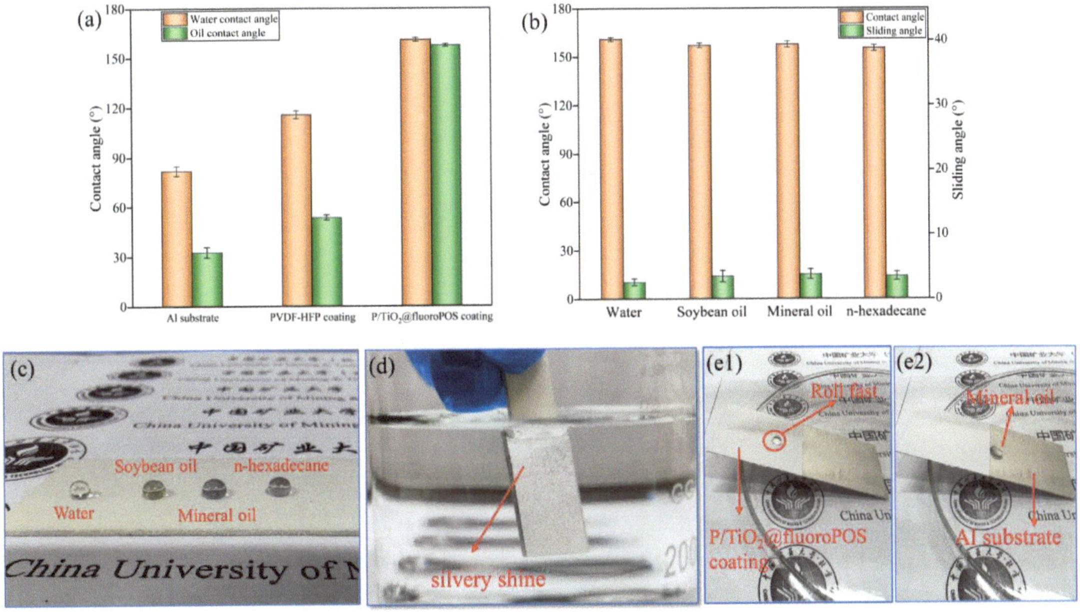

Figure 5. (**a**) Wettability of water and mineral oil on different samples. (**b**) The contact angle and sliding angle of various liquid droplets on P/TiO_2@fluoroPOS coating. (**c**) Photograph of various liquid droplets on the P/TiO_2@fluoroPOS coating. (**d**) The silver mirror effect. (**e1**,**e2**) Rolling of mineral oil on a substrate composed of P/TiO_2@fluoroPOS and Al.

The wettability of the P/TiO_2@fluoroPOS coatings was assessed by measuring the contact angles of water and mineral oil at various particle concentrations. The coatings maintained a filler-to-binder mass ratio of 4:3. The purpose was to investigate the impact of different particle concentrations on the wettability of the coatings and identify the critical concentration required to achieve superamphiphobic properties. Table 1 demonstrates a notable association between the concentration of particles and the hydrophobic and oleophobic characteristics of the coating. At a particle concentration of 6.19 g/m^2, the coating attains superhydrophobicity. When the particle concentration increases to 20.69 g/m^2, the oil contact angle of the coating measures 149.8°, which is near the critical concentration required for achieving superamphiphobicity. At a concentration of 29.97 g/m^2, the coating exhibits exceptional superamphiphobic properties. Subsequently, when the concentration increases further, the contact angle reaches a plateau.

Table 1. Influence of surface particle concentration on the liquid repellency of coatings.

Concentration of Particles (g/m^2)	Water Contact Angle (°)	Standard Deviation	Oil Contact Angle (°)	Standard Deviation
6.19	154.4	1.30115	136.3	2.15244
11.75	157.6	1.69794	144.1	2.61381
20.69	159.5	1.39104	149.8	1.56301
29.97	160.2	1.01833	154.8	1.57575
36.36	160.7	1.23369	157.2	1.10317

In order to evaluate the wettability of different surface tension (γ) liquids on the surface of the P/TiO_2@fluoroPOS coating, the contact angle and sliding angle of various droplets, such as water (γ = 72.8 mN/m), soybean oil (γ = 33.8 mN/m), mineral oil (γ = 30.7 mN/m), and n-hexadecane (γ = 27.4 mN/m) [30,40], were measured. All of these liquids had contact angles larger than 150° and sliding angles smaller than 10°, as seen in Figure 5b. The contact angle of water was measured to be 160.8°, whereas the contact angle of n-hexadecane was found to be 155.1°, and the sliding angle was measured to be 3.5°. The P/TiO_2@fluoroPOS coatings were coated with droplets of all four liquids, and, as shown in Figure 5c, the droplets were able to retain their spherical or ellipsoidal shapes. A silvery shine on the coating is observed when the coating is submerged in the liquid, as depicted in Figure 5d. This silvery shine was formed due to the total reflectance of light on the air film trapped on the surface. This specific event suggests that the coating's surface can isolate the coating from the liquid by trapping and holding a stable air film. When mineral oil drops are applied to two different surfaces of aluminum plates positioned at an incline (P/TiO_2@fluoroPOS coating, aluminum plate), as depicted in Figure 5(e1, e2), the droplets of mineral oil move quickly on the P/TiO_2@fluoroPOS coating without sticking to it. However, when the coating comes into contact with the aluminum plate, the droplets spread out rapidly, and the movement becomes more fluid, but at a slower pace. Additionally, the droplets adhere to the coating, demonstrating exceptional performance. The velocity was significantly reduced, and there were adhered oil droplets. The coatings of P/TiO_2@fluoroPOS exhibited exceptional superamphiphobic properties.

3.4. Corrosion Resistance of P/TiO_2@fluoroPOS Coatings

In order to evaluate the corrosion resistance and stability of the coatings in extreme external environments, they were immersed in solutions containing 2 mol/L of NaOH, HCl, and NaCl. Figure 6b demonstrates a clear silver mirror effect at the boundary between the solid and liquid. This indicates that the coating has successfully maintained a stable air layer in the corrosive solution environment. Following a 12-h immersion, the contact angle and sliding angle were determined. The results in Figure 6a revealed that the coating maintained a contact angle over 150° and a sliding angle below 10°, demonstrating exceptional superamphiphobic performance even after being exposed to the corrosive solution. The P/TiO_2@fluoroPOS coatings exhibit exceptional superamphiphobic characteristics and demonstrate a strong aversion to corrosive solutions. According to the Cassie state, air may be easily trapped in the small gaps between the protrusions. This creates a barrier that makes it challenging for water and corrosive ions to access the surface of the coating. The repulsion hinders the wetting of the coating surface by the liquid, consequently decreasing the contact area between the liquid and the coated surface. Corrosion typically happens when a liquid comes into contact with a material's surface and a chemical reaction occurs. By minimizing the contact area, corrosion can be effectively prevented.

3.5. Oil Repellency of P/TiO_2@fluoroPOS Coatings

Superamphiphobic coatings often exhibit antifouling characteristics, but with limitations imposed on their applicability to ordinary liquids such as water and cooking oil. However, mineral oil has a low surface tension and a high adherence to the substrate or coating surface; thus, the coating's resistance to adhesion must meet strict standards. To evaluate the durability of the coatings in an environment where they are immersed in mineral oil, specifically focusing on their superamphiphobic properties, the P/TiO_2@fluoroPOS coatings were immersed in mineral oil and then extracted, with each instance being documented as a cycle (Figure 6(d1)). Following a certain number of cycles, the coatings were assessed for their contact angles and slide angles. As shown in Figure 6c, after being immersed for 200 cycles, the coating continues to exhibit superamphiphobicity. However, after 400 cycles of immersion, the oil contact angle (OCA) of the coating reduces to 147.3°. The contact angle exhibits a gradual reduction as the number of cycles increases. As the cycles approached 1200, the contact angle consistently maintained over 135°, indicating a continued ability to effectively repel liquids. After an extended period of immersion cycle testing, small spherical droplets will adhere to the surface of the coating (Figure 6(d2,d3)). This is likely due to the infiltration of mineral oil into the micro-nano protrusions on the coating surface, which increases the adhesion force. Consequently, the coating transitions from the Cassie state to a coexisting state of Wenzel and Cassie [41,42]. During each measurement, when the mineral oil comes into contact with the coating surface, some of the micro-nano protrusions may be re-infiltrated, displacing air and causing a change in the contact angle.

The inclusion of PVDF-HFP and TiO_2@fluoroPOS fillers in the coating imparts it with low surface energy groups, resulting in an inert barrier and exceptional chemical resistance under challenging conditions, such as highly corrosive liquids and mineral oils. The combination of structural and compositional stability contributes to the overall stability of the collected gas film, ensuring that the coating offers versatile protection for the substrate in practical settings.

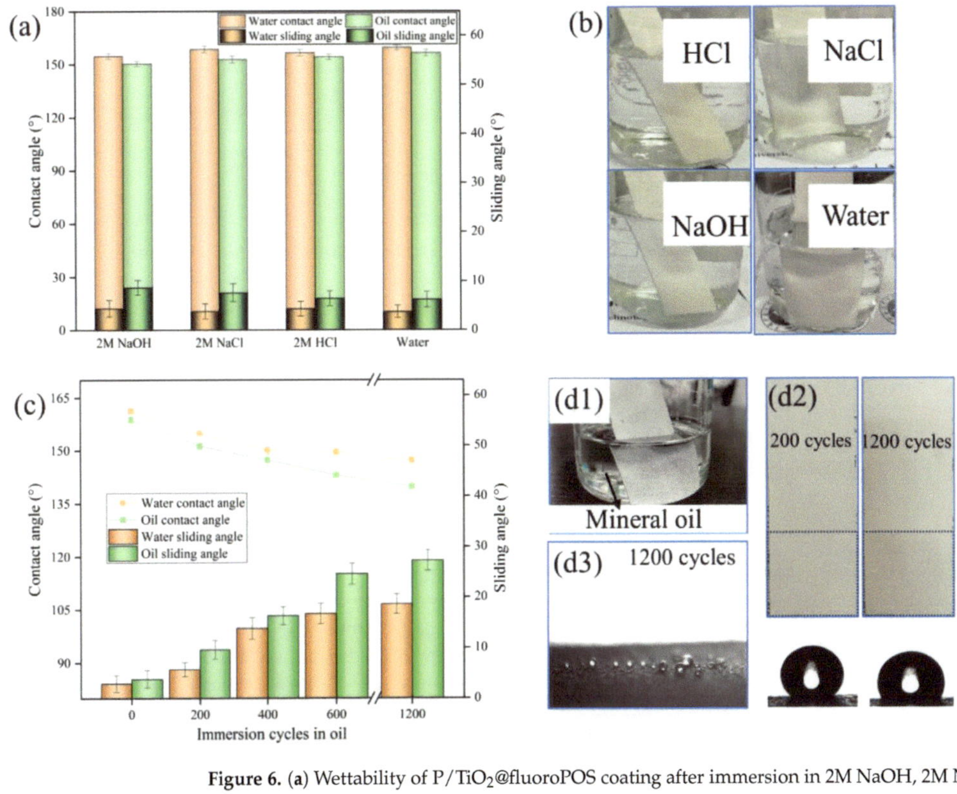

Figure 6. (**a**) Wettability of P/TiO_2@fluoroPOS coating after immersion in 2M NaOH, 2M NaCl, 2M HCl, and water for 12 h. (**b**) Photograph of P/TiO_2@fluoroPOS coating immersion in 2M NaOH, 2M NaCl, 2M HCl, and water. (**c**) Wettability of P/TiO_2@fluoroPOS coating with immersion cycles in mineral oil. (**d1**) Photograph of P/TiO_2@fluoroPOS coating immersion in mineral oil; (**d2**) photograph of the coating after 200 and 2000 cycles; (**d3**) droplets of oil sticking to the coating's surface.

3.6. Anti-Icing of P/TiO_2@fluoroPOS Coatings

Figure 7a,b illustrates the solidification of water on the Al plate's surface and the P/TiO_2@fluoroPOS coating at a temperature of −15 °C. The water droplets on the surface of the aluminum plate underwent full solidification after a duration of 530 s. By contrast, the water droplets on the P/TiO_2@fluoroPOS coating took 1254 s to freeze. The freezing duration of the water droplets was noticeably prolonged when shielded by the P/TiO_2@fluoroPOS coating. The inherent hydrophilicity of the uncoated aluminum plate facilitated the wetting of the coating surface by water droplets, leading to an increased contact area between the solid and liquid phases. As a consequence, heat transfer across the solid-liquid interface was accelerated at low temperatures. On the other hand, the shape of water droplets on the P/TiO_2@fluoroPOS coating is spherical, and the coating maintains its superhydrophobic properties even at low temperatures. The coating's surface is non-wetted, resulting in a significant decrease in the contact area between the solid and liquid. This results in a notable decrease in the rate at which heat is conducted, thereby causing a delay in the freezing of water droplets.

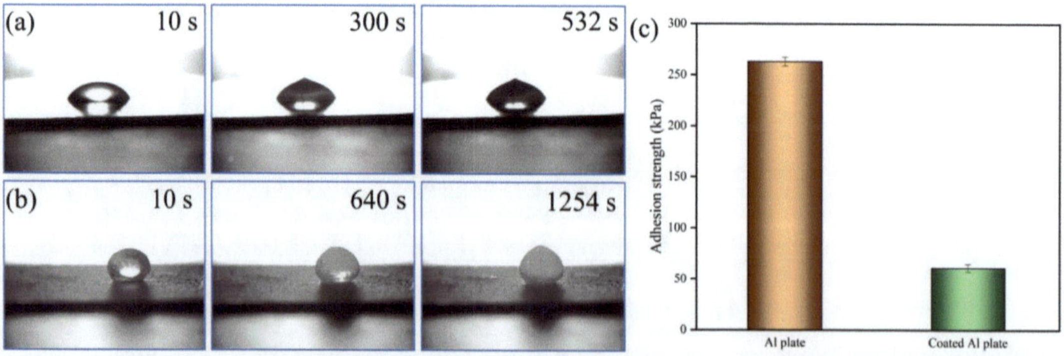

Figure 7. The freezing process of a water droplet (0.1 mL) at a temperature of −15 °C on the (**a**) Al plate and (**b**) P/TiO_2@fluoroPOS coating; (**c**) Ice adhesion strength on the Al plate and P/TiO_2@fluoroPOS coated Al plate at −15 °C.

At a temperature of −15 °C, the ice adhesion strength on the surface of an Al plate is 263.1 kPa, whereas the ice adhesion strength on the surface of a P/TiO_2@fluoroPOS coating is 61.2 kPa, as shown in Figure 7c. The ice adhesion strength on the coated surface is greatly reduced, facilitating the fast removal of ice from the surface. The coating may effectively prevent ice formation by delaying the freezing of water droplets and lowering the strength of ice adhesion, thereby achieving the objective of anti-icing.

3.7. *Anti-Waxing of P/TiO_2@fluoroPOS Coatings*

Wax accumulation issues in the petroleum industry may be resolved by using liquid-avoiding superamphiphobic coatings. Figure 8a,b illustrates the process of paraffin droplets forming wax on the surface of a carbon steel plate and P/TiO_2@fluoroPOS coating at a temperature of 20 °C. The paraffin droplets on the carbon steel plate underwent quick infiltration and dispersion, followed by full solidification within 440 milliseconds, resulting in strong adhesion to the iron plate's surface. On the other hand, the paraffin droplets present on the P/TiO_2@fluoroPOS coating exhibited an ellipsoidal shape and took 1240 ms to fully solidify. This considerably prolonged the time it took for wax to form. The wax droplets were applied to the coating surface and exhibited an ellipsoidal morphology. The contact area between the paraffin wax and the coating was minimal, suggesting that the coating had a propensity to evade the wax droplets. It was postulated that this phenomenon could be attributed to the presence of an air film on the coating surface, acting as an anti-adhesion layer, which impeded the wetting of the paraffin wax droplets. Simultaneously, the limited thermal conductivity of air impedes the propagation of heat from the paraffin droplets to the frigid surface. The presence of an air film serves as a thermal barrier and anti-adhesion layer, thereby impeding the pace of wax bonding and diminishing the bonding strength. Consequently, this coating proves to be efficacious in preventing wax bonding and effectively addresses the issue of anti-waxing during the transportation of crude oil.

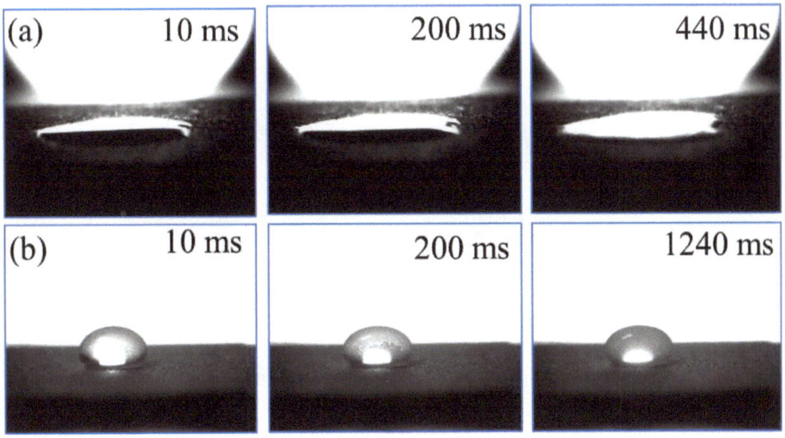

Figure 8. The waxing process of a paraffinic droplet (0.1 mL) at a temperature of 20 °C on the (**a**) carbon steel plate and (**b**) P/TiO_2@fluoroPOS coating.

3.8. Self-Cleaning Ability of P/TiO_2@fluoroPOS Coatings

Water and mineral oil were used as driving droplets to evaluate the self-cleaning efficacy of P/TiO_2@fluoroPOS coatings. As shown in Figure 9a–d, clays and coal were chosen as surface pollutants and equally distributed on the coated surface positioned at an incline. Drops of water and mineral oil were then introduced separately. The test findings indicated that the coating's surface facilitated the rolling movement of water and oil droplets, enabling them to penetrate and remove impurities from the coating's surface. The findings indicate the coating's exceptional ability to self-clean its surface in both water and oil liquid conditions.

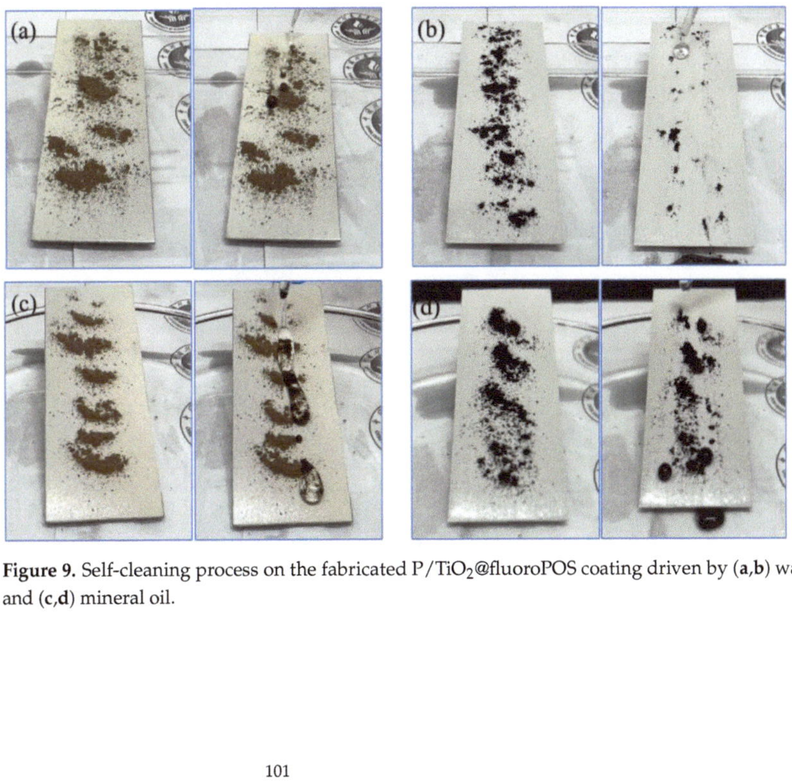

Figure 9. Self-cleaning process on the fabricated P/TiO_2@fluoroPOS coating driven by (**a**,**b**) water and (**c**,**d**) mineral oil.

3.9. Mechanical Stability of P/TiO_2@fluoroPOS Coatings

3.9.1. Water Jetting Test

Given the environmental conditions in which superamphiphobic coatings are used, fluid jetting is often inevitable. Water jetting experiments were performed to evaluate the impact resistance of the P/TiO_2@fluoroPOS coating. The coating was positioned in a horizontal orientation, 30 cm below the water flow, at a pressure of 80 kPa, as seen in Figure 10a. The contact angle and sliding angle of the coating were measured after a certain duration of water impact. The findings may be shown in Figure 10b. Following a 10-min water jetting process, the contact angle of the coating did not exhibit a substantial decline and remained over 150°. However, the sliding angle experienced a large rise, reaching x°. The armor structure of the TiO_2@fluoroPOS fillers successfully withstood the forceful flow of water, preventing quick degradation of the nano-structure of the coating and ultimately avoiding coating failure. During the time period of 10~18 min, when water is sprayed onto the coating, the contact angle decreases at a faster rate. This leads to a loss of superamphiphobicity, and some of the rough structure that is not strongly bonded with the binder is damaged. After 18 min of exposure to the water flow, the contact angle decreases at a slower rate and eventually stabilizes. Following a 30 min impact, the contact angle of the coating remained consistently over 130°.

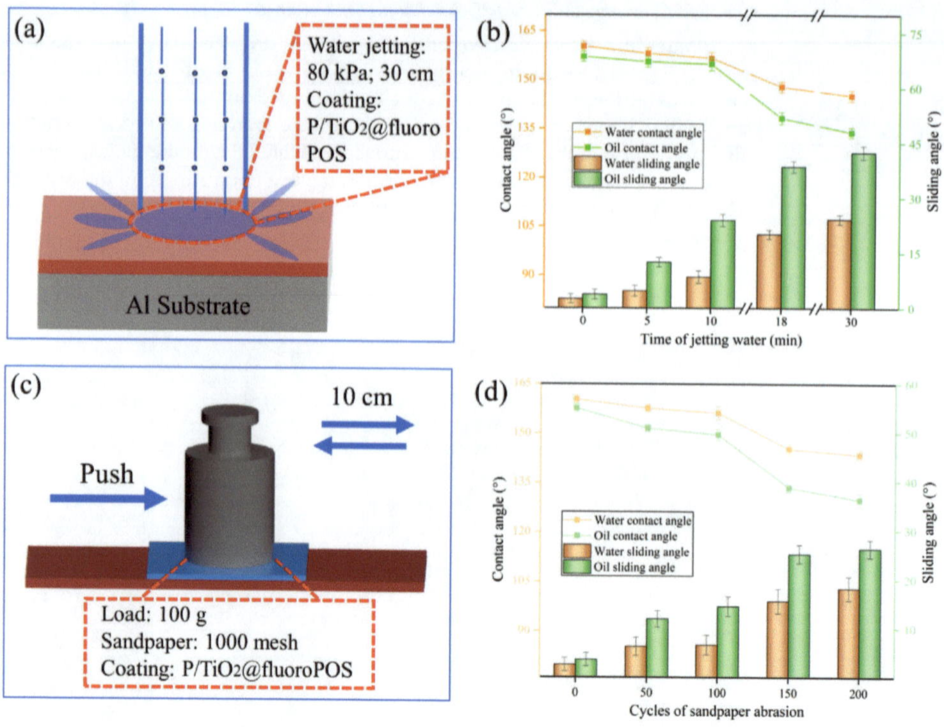

Figure 10. (**a**) Schematic of sandpaper abrasion; (**b**) the wettability of P/TiO_2@fluoroPOS coating after sandpaper abrasion. (**c**) Schematic of the water jetting test; (**d**) the wettability of the P/TiO_2@fluoroPOS coating after water jetting.

3.9.2. Sandpaper Abrasion Test

One of the main obstacles to the widespread use of superamphiphobic coatings is their surface structure's fragility. Sandpaper abrasion tests were used to assess the mechanical resilience of the P/TiO_2@fluoroPOS coatings that were produced. As shown in Figure 10c, square sandpaper was positioned squarely against the coating (contact area 16 cm^2) and moved 10 cm in a single direction while bearing a 100 g load set as a single cycle. Figure 10d displays the test findings. After 100 abrasion cycles, the coating's superamphiphobicity did not alter. After 200 cycles, however, a portion of the coating's surface structure was damaged, and its oca was still larger than 130°, maintaining strong liquid repellency. The nanoclusters attain a micron-scale size subsequent to the development of protective armor on their surface. The micron particles undergo a secondary rough structure on the surface, resulting in a nano-meter scale structure. The binder secures the micron particles to withstand tangential forces, while the surface armor safeguards the nano-particles from damage to the nano-structure. The combined effect of the binder and protective armor ensures exceptional mechanical stability for the coating.

3.10. Superamphiphobicity Mechanism

PVDF-HFP and TiO_2@fluoroPOS fillers enhanced with low surface energy groups—which have outstanding liquid repellency, corrosion resistance, and mechanical stability—make up the P/TiO_2@fluoroPOS superamphiphobic coating. As seen in Figure 11a, a potential mechanism for the interaction between the coating and liquid was suggested in order to clarify the superamphiphobicity mechanism of the coating. Many micro-convex and inwardly concave microporous structures were formed by the accumulation of fillers on the coating surface. By using this structure and the low surface energy TiO_2@fluoroPOS fillers, an air film could be captured on the coating's surface. The nano-microstructures on the coating's surface are a part of the fillers' rough armor, which is difficult to destroy. The rough armor is fixed by the PVDF-HFP connection between the encapsulated agglomerates, ensuring stability for both the structure and the captured air film. The Cassie–Baxter equation [17,21,43] states that $\cos\theta_D = f_s\cos\theta_Y - f_g$, where θ_D represents the apparent contact angle and θ_Y represents the liquid-solid intrinsic contact angle. In the ideal Cassie state, the sum of f_s and f_g (liquid-solid/gas-solid contact fractions) is 1, resulting in $\cos\theta_D = f_s(1 + \cos\theta_Y) - 1$. Therefore, as the liquid-solid contact fraction fs decreases, the apparent contact angle θ_D significantly increases, leading to the repulsion of low surface energy droplets.

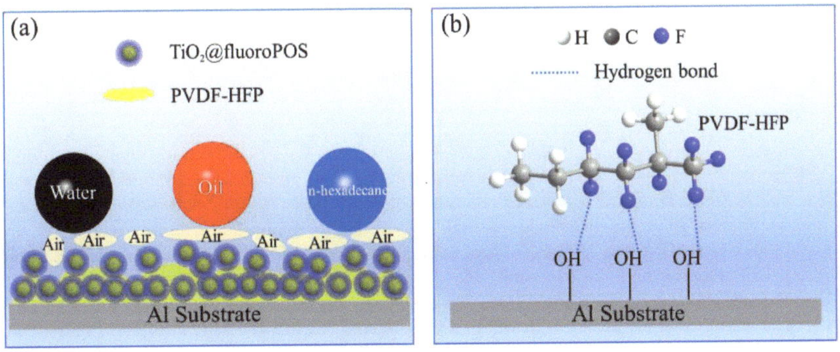

Figure 11. (**a**) Schematic diagram of the mechanisms of air film stability on the P/TiO_2@fluoroPOS coating surface. (**b**) Hydrogen bond interactions between the Al substrate and PVDF-HFP.

PVDF-HFP serves as a binding agent in the coating process, facilitating improved interparticle bonding and promoting robust adhesion to the coating. This enables a secure link between the substrate and the coating, thereby mitigating tangential friction. Figure 11b illustrates the hydrogen bonding interactions between the F atoms in the PVDF-HFP and the -OH groups on the Al substrate.

4. Conclusions

In this study, a novel superamphiphobic coating consisting of P/TiO_2@fluoroPOS was effectively synthesized using a straightforward and scalable spraying method. The coating exhibited corrosion resistance, self-cleaning properties, anti-fouling capabilities, anti-ice-covering properties, and anti-waxing properties. The superamphiphobic filler with micro-nano structure was obtained by growing rough armors on the surface of titanium dioxide agglomerates using the sol-gel process and subsequently fluorinating them. The coatings constructed using PVDF-HFP and TiO_2@fluoroPOS fillers exhibit highly stable air films due to the synergistic effects of filler structural stabilization and binder immobilization.

Many liquids with varying surface tensions exhibit repulsion towards the coating, including n-hexadecane (contact angle: 155.1°, sliding angle: 3.5°) and water (160.8°, sliding angle: 2.6°). Furthermore, the coating exhibits remarkable mechanical and chemical stability, and it retains its superamphiphobicity even after being immersed for 12 h in solutions containing 2 mol/L HCl, NaOH, and NaCl. In addition, the coating oil contact angle retains more than 130° even after 200 cycles of sandpaper abrasion and 30 min of 80 kPa water jetting. The coating also performs exceptionally well in terms of self-cleaning, anti-icing, and anti-waxing. With a wide range of potential applications in real-world industrial production, the current work offers a novel concept for immobilizing nanoparticles on the coating's surface to improve its mechanical and chemical resilience.

Author Contributions: Conceptualization, X.H. and X.G.; methodology, X.H., X.G. and S.Z.; writing—original draft preparation, X.H. and X.G.; writing—review and editing, X.H., X.G., X.W., H.S. and S.Z.; supervision, X.H., X.G., X.W. and H.S.; funding acquisition, X.H., S.Z. and H.S. All authors have read and agreed to the published version of the manuscript.

Funding: This research was funded by the National Science and Technology Major Project (J2019-VII-0015-0155) and the Fundamental Research Funds for the Central Universities (grant number 2023ZKPYJD01).

Institutional Review Board Statement: Not applicable.

Informed Consent Statement: Not applicable.

Data Availability Statement: The raw data supporting the conclusions of this article will be made available by the authors on request.

Conflicts of Interest: The authors declare no conflicts of interest.

References

1. Ren, G.; Qiao, Z.; Hui, Z.; Tuo, Y.; Zheng, W.; Chen, X.; Li, S. The Waterborne Superamphiphobic Coatings with Antifouling, High Temperature Resistance, and Corrosion Resistance. *ACS Omega* **2023**, *8*, 13578–13592. [CrossRef] [PubMed]
2. Shi, T.; Wang, H.; Jia, Q.; Chen, R. Preparation of a transparent coating with superamphiphobic and antifouling properties. *Mater. Chem. Phys.* **2023**, *293*, 126888. [CrossRef]
3. Qiao, Z.; Ren, G.; Chen, X.; Gao, Y.; Tuo, Y.; Lu, C. Fabrication of Robust Waterborne Superamphiphobic Coatings with Antifouling, Heat Insulation, and Anticorrosion. *ACS Omega* **2022**, *8*, 804–818. [CrossRef]
4. Yuan, R.; Wu, S.; Yu, P.; Wang, B.; Mu, L.; Zhang, X.; Zhu, Y.; Wang, B.; Wang, H.; Zhu, J. Superamphiphobic and Electroactive Nanocomposite toward Self-Cleaning, Antiwear, and Anticorrosion Coatings. *ACS Appl. Mater. Interfaces* **2016**, *8*, 12481–12493. [CrossRef] [PubMed]
5. Jiao, L.; Xu, Q.; Tong, J.; Liu, S.; Hu, Y.; Guo, Q.; Wu, H.; Li, W.; Zhao, Q.; Chen, R. Facile preparation of pliable superamphiphobic papers with high and durable liquid repellency for anti-corrosion and open surface microfluidics. *Appl. Surf. Sci.* **2022**, *606*, 154845, Erratum in *Appl. Surf. Sci.* **2023**, *617*, 155517. [CrossRef]
6. Zhang, B.; Yang, G.; Xu, W.; Duan, J.; Hou, B. Hybrid superamphiphobic anti-corrosion coating with integrated functionalities of liquid repellency, self-cleaning, and anti-icing. *J. Mater. Sci. Technol.* **2024**, *184*, 256–268. [CrossRef]
7. Yang, Y.; Zhang, Y.; Li, G.; Zhang, M.; Wang, X.; Song, Y.; Liu, S.; Cai, Y.; Wu, D.; Chu, J.; et al. Directional rebound of compound droplets on asymmetric self-grown tilted mushroom-like micropillars for anti-bacterial and anti-icing applications. *Chem. Eng. J.* **2023**, *472*, 144949. [CrossRef]
8. Wei, J.; Liang, W.; Zhang, J. Preparation of Mechanically Stable Superamphiphobic Coatings via Combining Phase Separation of Adhesive and Fluorinated SiO2 for Anti-Icing. *Nanomaterials* **2023**, *13*, 1872. [CrossRef] [PubMed]

9. Wang, H.; Lu, H.; Zhang, X. Super-robust superamphiphobic surface with anti-icing property. *RSC Adv.* **2019**, *9*, 27702–27709. [CrossRef]
10. Long, Y.; Yin, X.; Mu, P.; Wang, Q.; Hu, J.; Li, J. Slippery liquid-infused porous surface (SLIPS) with superior liquid repellency, anti-corrosion, anti-icing and intensified durability for protecting substrates. *Chem. Eng. J.* **2020**, *401*, 126137. [CrossRef]
11. Li, Z.; Li, K.; Li, X.; Feng, Y.; Li, H.; Wang, H. Preparation of linseed oil-loaded porous glass bubble/wax microcapsules for corrosion- and wear-resistant difunctional coatings. *Chem. Eng. J.* **2022**, *437*, 135403. [CrossRef]
12. Meena Narayana Menon, D.; Giardino, M.; Janner, D. Tunable pulsewidth nanosecond laser texturing: From environment friendly superhydrophobic to superamphiphobic surfaces. *Appl. Surf. Sci.* **2023**, *610*, 155356. [CrossRef]
13. Karapanagiotis, I.; Manoudis, P.N. Superhydrophobic and superamphiphobic materials for the conservation of natural stone: An overview. *Constr. Build. Mater.* **2022**, *320*, 126175. [CrossRef]
14. Cao, C.; Yi, B.; Zhang, J.; Hou, C.; Wang, Z.; Lu, G.; Huang, X.; Yao, X. Sprayable superhydrophobic coating with high processibility and rapid damage-healing nature. *Chem. Eng. J.* **2020**, *392*, 124834. [CrossRef]
15. Yang, Y.; He, H.; Li, Y.; Qiu, J. Using Nanoimprint Lithography to Create Robust, Buoyant, Superhydrophobic PVB/SiO_2 Coatings on wood Surfaces Inspired by Red roses petal. *Sci. Rep.* **2019**, *9*, 9961. [CrossRef] [PubMed]
16. Zhang, Z.-h.; Wang, H.-j.; Liang, Y.-h.; Li, X.-j.; Ren, L.-q.; Cui, Z.-q.; Luo, C. One-step fabrication of robust superhydrophobic and superoleophilic surfaces with self-cleaning and oil/water separation function. *Sci. Rep.* **2018**, *8*, 3869. [CrossRef] [PubMed]
17. Ellinas, K.; Tserepi, A.; Gogolides, E. Durable superhydrophobic and superamphiphobic polymeric surfaces and their applications: A review. *Adv. Colloid Interface Sci.* **2017**, *250*, 132–157. [CrossRef] [PubMed]
18. Chen, L.; Guo, Z.; Liu, W. Biomimetic Multi-Functional Superamphiphobic FOTS-TiO2 Particles beyond Lotus Leaf. *ACS Appl. Mater. Interfaces* **2016**, *8*, 27188–27198. [CrossRef]
19. Zhou, H.; Niu, H.; Wang, H.; Lin, T. Self-Healing Superwetting Surfaces, Their Fabrications, and Properties. *Chem. Rev.* **2022**, *123*, 663–700. [CrossRef]
20. Si, W.; Guo, Z. Enhancing the lifespan and durability of superamphiphobic surfaces for potential industrial applications: A review. *Adv. Colloid Interface Sci.* **2022**, *310*, 102797. [CrossRef]
21. Singh, S.; Kango, S.; Sharma, N.; Verma, R. Recent advances in the mechanical durability of superamphiphobic surfaces: A review. *Proc. Inst. Mech. Eng. Part J J. Eng. Tribol.* **2021**, *235*, 2474–2499. [CrossRef]
22. Bao, M.; Tie, L.; Li, J. Smart superamphiphobic surface manipulating wetting behaviors of oil droplet in water. *Tribol. Int.* **2023**, *179*, 108189. [CrossRef]
23. Hegner, K.I.; Hinduja, C.; Butt, H.-J.; Vollmer, D. Fluorine-Free Super-Liquid-Repellent Surfaces: Pushing the Limits of PDMS. *Nano Lett.* **2023**, *23*, 3116–3121. [CrossRef] [PubMed]
24. Dong, J.; Zhang, J. Biomimetic Super Anti-Wetting Coatings from Natural Materials: Superamphiphobic Coatings Based on Nanoclays. *Sci. Rep.* **2018**, *8*, 12062. [CrossRef]
25. Zhang, H.; Xue, P.; Wang, M.; Wang, J.; Shi, Y.; Pan, L. Fabrication of superamphiphobic surface with re-entrant structures via self-assembly colloidal template-assisted electrochemical deposition. *Surf. Interfaces* **2023**, *40*, 103033. [CrossRef]
26. Vu, H.H.; Nguyen, N.T.; Kashaninejad, N. Re-Entrant Microstructures for Robust Liquid Repellent Surfaces. *Adv. Mater. Technol.* **2023**, *8*, 2201836. [CrossRef]
27. Kang, S.M.; Choi, J.S. Selective Liquid Sliding Surfaces with Springtail-Inspired Concave Mushroom-Like Micropillar Arrays. *Small* **2019**, *16*, 1904612. [CrossRef] [PubMed]
28. Ganesh, V.A.; Dinachali, S.S.; Nair, A.S.; Ramakrishna, S. Robust Superamphiphobic Film from Electrospun TiO_2 Nanostructures. *ACS Appl. Mater. Interfaces* **2013**, *5*, 1527–1532. [CrossRef] [PubMed]
29. Abu Jarad, N.; Imran, H.; Imani, S.M.; Didar, T.F.; Soleymani, L. Fabrication of Superamphiphobic Surfaces via Spray Coating; a Review. *Adv. Mater. Technol.* **2022**, *7*, 2101702. [CrossRef]
30. Zhu, P.; Zhu, L.; Ge, F.; Wang, G.; Zeng, Z. Robust and transparent superamphiphobic coating prepared via layer-by-layer spraying. *Surf. Coat. Technol.* **2021**, *426*, 127793. [CrossRef]
31. Zhang, B.; Xu, W.; Xia, D.; Huang, Y.; Zhao, X.; Zhang, J. Spray coated superamphiphobic surface with hot water repellency and durable corrosion resistance. *Colloids Surf. A Physicochem. Eng. Asp.* **2020**, *596*, 124750. [CrossRef]
32. Liu, X.; Wang, K.; Zhang, W.; Zhang, J.; Li, J. Robust, self-cleaning, anti-fouling, superamphiphobic soy protein isolate composite films using spray-coating technique with fluorinated HNTs/SiO_2. *Compos. Part B Eng.* **2019**, *174*, 107002. [CrossRef]
33. Zhou, X.; Liu, J.; Liu, W.; Steffen, W.; Butt, H.J. Fabrication of Stretchable Superamphiphobic Surfaces with Deformation-Induced Rearrangeable Structures. *Adv. Mater.* **2022**, *34*, 2107901. [CrossRef] [PubMed]
34. Si, W.; Wu, J.; Guo, Z. A simple preparation of a F-TiO_2-HNT superamphiphobic surface with a tube-point-like micro/nano hierarchical structure for self-cleaning and anti-fouling. *New J. Chem.* **2023**, *47*, 9989–9993. [CrossRef]
35. Wang, T.; Lv, C.; Ji, L.; He, X.; Wang, S. Designing Re-Entrant Geometry: Construction of a Superamphiphobic Surface with Large-Sized Particles. *ACS Appl. Mater. Interfaces* **2020**, *12*, 49155–49164. [CrossRef] [PubMed]
36. Zhou, X.; Sun, S.; Zhang, C.; Wang, X.-Y.; Li, Y.-L.; Jiang, Y. Facile fabrication of durable superamphiphobic PET fabrics. *J. Coat. Technol. Res.* **2019**, *17*, 711–718. [CrossRef]
37. Li, X.; Li, H.; Huang, K.; Zou, H.; Dengguang, Y.; Li, Y.; Qiu, B.; Wang, X. Durable superamphiphobic nano-silica/epoxy composite coating via coaxial electrospraying method. *Appl. Surf. Sci.* **2018**, *436*, 283–292. [CrossRef]

38. Peng, J.; Yuan, S.; Geng, H.; Zhang, X.; Zhang, M.; Xu, F.; Lin, D.; Gao, Y.; Wang, H. Robust and multifunctional superamphiphobic coating toward effective anti-adhesion. *Chem. Eng. J.* **2022**, *428*, 131162. [CrossRef]
39. Kuna, J.J.; Voïtchovsky, K.; Singh, C.; Jiang, H.; Mwenifumbo, S.; Ghorai, P.K.; Stevens, M.M.; Glotzer, S.C.; Stellacci, F. The effect of nanometre-scale structure on interfacial energy. *Nat. Mater.* **2009**, *8*, 837–842. [CrossRef]
40. Masoud Emarati, S.; Mozammel, M. Theoretical, fundamental and experimental study of Liquid-repellency and corrosion resistance of fabricated superamphiphobic surface on Al alloy 2024. *Chem. Eng. J.* **2020**, *387*, 124046. [CrossRef]
41. Tuteja, A.; Choi, W.; Ma, M.; Mabry, J.M.; Mazzella, S.A.; Rutledge, G.C.; McKinley, G.H.; Cohen, R.E. Designing Superoleophobic Surfaces. *Science* **2007**, *318*, 1618–1622. [CrossRef] [PubMed]
42. Wang, S.; Jiang, L. Definition of Superhydrophobic States. *Adv. Mater.* **2007**, *19*, 3423–3424. [CrossRef]
43. Milionis, A.; Bayer, I.S.; Loth, E. Recent advances in oil-repellent surfaces. *Int. Mater. Rev.* **2016**, *61*, 101–126. [CrossRef]

materials

MDPI

Article

Investigation of Grain Boundary Effects in $Sm_{0.2}Ce_{0.8}O_{2-x}$ Thin Film Memristors

Weikai Shi, Luyao Wang and Nan Yang *

School of Physical Science and Technology, ShanghaiTech University, Shanghai 201210, China; shiwk@shanghaitech.edu.cn (W.S.); wangly2@shanghaitech.edu.cn (L.W.)
* Correspondence: yangnan@shanghaitech.edu.cn

Abstract: Cerium-based materials (CeO_{2-x}) are of significant interest in the development of vacancy-modulated resistive switching (RS) memory devices. However, the influence of grain boundaries on the performance of memristors is very limited. To fill this gap, this study explores the influence of grain boundaries in cerium-based thin film resistive random-access memory (RRAM) devices. $Sm_{0.2}Ce_{0.8}O_{2-x}$ (SDC20) thin films were deposited on (100)-oriented Nb-doped $SrTiO_3$ (NSTO) and (110)-oriented NSTO substrates using pulsed laser deposition (PLD). Devices constructed with a Pt/SDC20/NSTO structure exhibited reversible and stable bipolar resistive switching (RS) behavior. The differences in conduction mechanisms between single-crystal and polycrystalline devices were confirmed, with single-crystal devices displaying a larger resistance window and higher stability. Combining the results of XPS and I–V curve fitting, it was confirmed that defects near the grain boundaries in the SDC-based memristors capture electrons, thereby affecting the overall performance of the RRAM devices.

Keywords: memristor; $Sm_{0.2}Ce_{0.8}O_{2-x}$; pulse laser deposition; grain boundaries; oxygen vacancy

Citation: Shi, W.; Wang, L.; Yang, N. Investigation of Grain Boundary Effects in $Sm_{0.2}Ce_{0.8}O_{2-x}$ Thin Film Memristors. *Materials* **2024**, *17*, 3360. https://doi.org/10.3390/ma17133360

Academic Editor: Daniel Georgiev

Received: 4 June 2024
Revised: 25 June 2024
Accepted: 3 July 2024
Published: 8 July 2024

1. Introduction

With the emergence of the Internet of Things (IoT) and cloud computing, there has been an exponential growth in data volume [1]. However, the classical computing architecture based on the von Neumann model exhibits fundamental limitations due to the so-called von Neumann bottleneck, which prevents simultaneous processing of data [2–4]. Memristors, as a type of non-volatile memory device based on RS, are regarded as an effective solution to overcome this issue due to their advantages, such as long data retention time, simple structure, high density integration, low power consumption, fast operation speed, strong scalability, straightforward components, and ease of integration with standard metal–oxide–semiconductor (MOS) technology [5]. Oxides are widely used in various fields due to their advantages, such as room temperature stability, multifunctionality, renewability, and processability [6]. In recent years, numerous oxide materials have been discovered to exhibit RS behavior, including SiO_2 [7], TiO_2 [8], Ta_2O_5 [9], HfO_2 [10], $BiFeO_3$ [11], and $SrTiO_3$ [12], among others. In general, the switching of memristors is caused by the formation and disruption of conductive filaments, which can typically be classified into the following two cases: metal cation pathways and oxygen vacancy anion pathways [1].

CeO_{2-x}, a classic fluorite-structured oxide, is widely investigated due to its considerable oxygen ion conductivity, structural stability, and other properties [13–15]. Introducing trivalent ions dopant can induce charge compensation defects in CeO_2, such as oxygen vacancies and Ce^{3+}, thereby enhancing the conductivity and oxygen storage properties of doped cerium ($Do_xCe_{1-x}O_{2-y}$) [16]. Although the formation of conductive filaments in memristors based on the VCM (Valence Change Memory) mechanism is not yet fully understood, most studies indicate a close association with oxygen vacancies [17–19]. Consequently, $Do_xCe_{1-x}O_{2-y}$ (e.g., SDC, GDC) is a potential material for the RS layer in RRAM.

Recently, extensive studies have been conducted on thin film CeO_{2-x} as the RS layer. Ismail et al. studied the RS characteristics of Ti/CeO_{2-x}/Pt devices with varying RS layer thicknesses and found that the electroforming voltage and set voltage decrease with decreasing film thickness. Rupp et al. discovered that Pt/$Gd_xCe_{1-x}O_{2-y}$/Pt devices with a doping concentration of 20 mol% exhibit optimal RS performance, while too low or too high doping concentrations have negative effects [20,21]. In CeO_{2-x}, in addition to oxygen vacancies, the impact of planar defects, such as grain boundaries, are also crucial, as they significantly influence catalytic performance, electrical conductivity, and the oxygen storage and release capabilities [22,23]. However, the grain boundary effects in SDC thin film memristors have not been extensively investigated. Therefore, to fill this gap, we fabricated single-crystalline and polycrystalline SDC thin film memristors and analyzed the effects of grain boundaries in these devices.

Pulsed laser deposition (PLD) provides an effective method for the precise control of the microstructure of target materials, taking device miniaturization into consideration. Recent studies have utilized PLD for the microstructural manipulation of target materials. For instance, Dou et al. utilized temperature engineering to achieve three distinct grain morphologies in CeO_2 thin films and demonstrated that devices with a "columnar scaffold" morphology exhibited the best overall resistive switching performance [24]. Additionally, Wang et al. used PLD to grow CeO_2(111) unit cells on STO(110) substrates, which have a high lattice mismatch, resulting in CeO_2 thin films with a significant number of grain boundaries [25]. Therefore, PLD is an effective method for defect engineering in cerium dioxide thin films, capable of introducing planar defects within the films. This provides an ideal platform for investigating the impact of grain boundaries on memristors.

This paper aims to explore the impact of grain boundaries on SDC thin film memristors. Using PLD, we grew both single-crystal and polycrystalline SDC films on (100)-oriented NSTO and (110)-oriented NSTO substrates, respectively. We investigated the bipolar resistive switching characteristics of devices with a Pt/SDC/NSTO structure, finding that single-crystal SDC film devices exhibited superior switching characteristics. Additionally, combining XPS and I–V curve fitting confirmed the differences in electron transport properties between single-crystalline and polycrystalline devices. Although atomic vacancies [26] enable CeO_2 to perform resistive switching, grain boundaries have a negative impact. This evidence is important for defect engineering and provides new insights for the development of advanced doped cerium-based RRAM devices in the future.

2. Materials and Methods

2.1. Synthesis of SDC20 Powder

CeO_2 and Sm_2O_3 powders were placed into a ball milling jar at the stoichiometric ratio, along with an appropriate amount of water as a dispersant, leaving approximately one-third of the space empty. Subsequently, an appropriate amount of ZrO_2 spheres was added as the grinding media. Two similarly weighted ball milling jars were then placed into a ball miller with a rotation speed set at 30 Hz and milled for 12 h. Upon completion of milling, the resulting precursor powder was dried in an oven. The dried powder was then placed in a muffle furnace and calcined at 1000 °C in air for 5 h at a heating rate of 5 °C/min.

2.2. Thin Film Preparation of the Resistive Switching Devices

Single-crystal (100)- and (110)-oriented $SrTiO_3$ substrates doped with 0.7 wt.% Nb, measuring 5×5 mm^2, were subjected to ultrasonic cleaning in acetone and ethanol prior to loading into the UHV (ultra-high vacuum) chamber. Subsequently, SDC20 thin films were deposited on the substrates via pulsed laser deposition (NanoPLD, PVD Products, Wilmington, MA, USA) under a base pressure lower than 5×10^{-7} Pa and without any gas environment. SDC20 targets were prepared using conventional sintering methods. Plasma was generated between the substrate and target using a focused KrF excimer laser (wavelength: 248 nm) with a laser energy of 0.5 J/cm^2 and a repetition rate of 10 Hz. Prior to SDC20 deposition, the substrate underwent in situ annealing at 650 °C to achieve a clean and flat surface. During the deposition process, the substrate was maintained at 650 °C. After deposition, the temperature was decreased to room temperature at a rate of 30 °C/min.

2.3. Resistive Switching Device Fabrication

Platinum (Pt) top electrodes (TE) were deposited at room temperature using radio frequency magnetron sputtering (Lesker ProlinePVD75, Kurt J. Lesker, Pittsburgh, PA, USA). The area of the TE was 100 × 100 μm^2, and a mask was employed for patterning. Additionally, Pt was also used to contact the NSTO bottom electrode (BE).

2.4. Electrical Characterization of Resistive Switching Device

In this study, all electrical characterization measurements were conducted using an Agilent B1500 semiconductor parameter analyzer. Measurements were performed by contacting the electrodes using probes, with the top electrode biased positively and the bottom electrode grounded.

3. Results and Discussion

3.1. Powder Characterization and Analysis

The powder XRD analysis, as shown in Figure S1, reveals the formation of a cubic fluorite structure in the powder, with no apparent impurity phases. A comparison with the pure CeO_2 structure indicates a slight increase in the 2θ diffraction peak of CeO_2(220) from 47.46° to 47.67°, accompanied by a reduction in the c-axis lattice parameter. This reduction is attributed to the substitution of Sm^{3+} (0.958 Å) for Ce^{4+} (0.97 Å), leading to lattice contraction, consistent with the expected results. The doping of 20% Sm was further confirmed by Inductively Coupled Plasma Spectrometer (ICP) analysis of the powder composition, as shown in Table S1.

3.2. Characterization and Analysis of SDC20 Thin Films

Figure 1a shows the X-ray diffraction (XRD) θ-2θ scan spectrum of the SDC20 film grown on (100)-oriented NSTO. According to the XRD spectra, only reflections corresponding to the (200) planes were observed on the SDC20 film, indicating that the SDC20 film is predominantly oriented along the c-axis. Figure 1d shows the X-ray diffraction (XRD) θ–2θ scan spectrum of the SDC20 film grown on (110)-oriented NSTO. Using Bragg's formula, we determined the lattice parameters of single-crystalline and polycrystalline SDC thin films along the c-axis to be 5.44 Å and 5.46 Å, respectively. The greater distortion along the c-axis in polycrystalline SDC thin films indicates the presence of more defects within the bulk of the polycrystalline films. Figure 1b,e depict the rocking curves of the single-crystal and polycrystalline SDC20 films, respectively. The half-widths are in the range of 0.04°~0.06°, indicating that both possess high crystal quality. Figures 1c and 1f, respectively show the two-dimensional AFM images of the single-crystal and polycrystalline SDC20 thin films. Both exhibit dense and smooth surfaces with uniform grain sizes and orderly arrangements. The AFM results show that the surface roughness of single-crystal and polycrystalline films is essentially the same, indicating that the contact between the Pt electrode and SDC is consistent in both single-crystal and polycrystalline devices. As shown in Figure S2, the surface morphology of single-crystal and polycrystalline SDC20 films was observed under a magnification of 50,000 times. Consistent with the AFM results presented in the previous section, both exhibited smooth and dense surfaces, indicating good film growth.

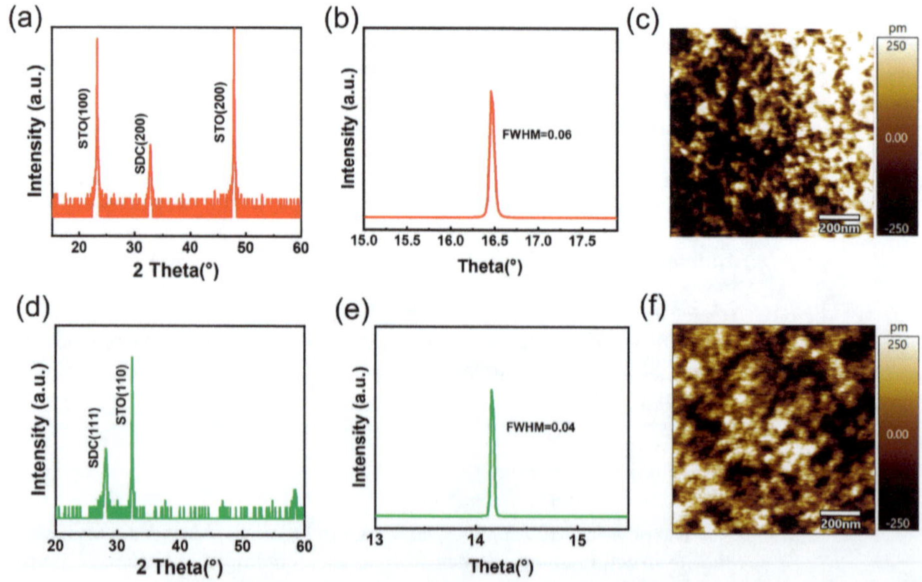

Figure 1. X-ray Diffraction (XRD) Spectrum and Atomic Force Microscopy (AFM) Images of Single-Crystal and Polycrystalline Devices: (**a**,**d**) high-resolution XRD spectrum of the SDC20 thin film. (**b**,**e**) Swinging curve at the (200) peak position of the SDC20 thin film. (**c**,**f**) Two-dimensional AFM image of the SDC20 thin film.

3.3. Electrical Characterization of the Devices

Figure 2a illustrates the current—voltage (I–V) characteristics of Pt/SDC20/NSTO(100) devices measured in voltage sweep mode. Typical bipolar resistive switching (RS) behavior is observed over repeated switching cycles. Figure 2a provides a brief description of the device structure, wherein forward bias corresponds to the application of positive voltage to the top Pt electrode. Furthermore, our Pt/SDC20/NSTO(100) structure does not require a forming process. As shown in Figure 2b, the initial resistance of the Pt/SDC20/NSTO(100) device remains in a high-resistance state (HRS). When the voltage is swept from 0 V to +4 V and back to 0 V, hysteresis in the I–V curve is observed, indicating that the device has been set to a low-resistance state (LRS), i.e., the set process. By applying a negative bias, the device can return to HRS, as indicated in ranges 3 and 4 of Figure 2b, i.e., the reset process. Obtaining identical I–V hysteresis over 50 consecutive cycles demonstrates reversible and repeatable RS behavior. Figure 2c illustrates the multi-step set process of the device. By consecutively applying positive scanning voltages of 5 V for six cycles, the current gradually increases with each step of the set process, demonstrating analog memristive device characteristics. This observation is further corroborated by subsequent synaptic plasticity tests, as depicted in Figure 2d. Under a read voltage of 0.5 V, the conductivity changes were sequentially recorded during 50 identical positive pulses (5 V, 5 ms) and 50 identical negative pulses (−5 V, 5 ms). The conductivity of the device gradually increased with the application of positive pulses, reaching saturation thereafter. In contrast, negative pulses led to a decrease in conductivity until saturation. These nonlinear transmission pulse-time-dependent characteristics bear a similarity to the pulse-time-dependent plasticity observed in biological synapses. As depicted in Figure 2e, the resistance ratio (R_{HRS}/R_{LRS}) between HRS and LRS at +0.5 V (~180) indicates the applicability of our RRAM device. Throughout over 500 switching cycles, no significant degradation or drastic fluctuations in the bistable resistance states were observed. Additionally, following the switching of the device to either the High Resistance State (HRS) or the Low Resistance State (LRS), the time retention capabilities of the device in both HRS and LRS were assessed under a reading voltage of 0.5 V applied every 10 s (Figure 2f). The resistance values of both HRS and LRS remained stable, with no decline observed within 10^4 s, thus confirming the non-volatile nature of the device. Similarly, we also conducted electrical performance tests on the polycrystalline Pt/SDC20/NSTO(110) devices. Figure 2g plots the current−voltage (I–V) characteristics of the Pt/SDC20/NSTO(110) device measured in voltage sweep mode, displaying bipolar I–V hysteresis loops within the range of −10 V to +7 V. As shown in Figure 2h, unlike the single-crystal device, after 100 cycles at a read voltage of 0.5 V, the R_{on}/R_{off} ratio is only 3. Similarly, as shown in Figure 2i, the polycrystalline devices also exhibited analog memristive characteristics, demonstrating potentiation and depression under 100 consecutive positive pulses and 100 consecutive negative pulses, respectively.

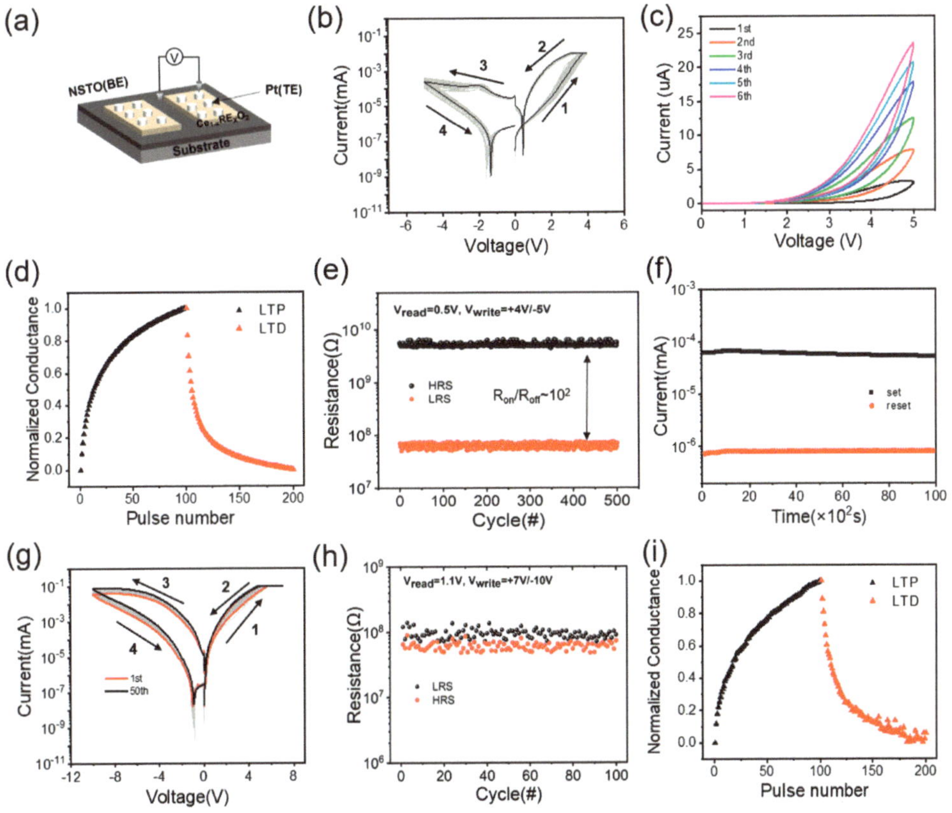

Figure 2. Electrical performance characterization of single-crystal (**a**–**f**) and polycrystalline (**g**–**i**) devices: (**a**) Schematic diagram of the device structure. (**b**) Typical I–V characteristic curves over 50 repeated switching cycles. (**c**) Multi-step set process of the device under a +5 V scanning voltage. (**d**) Synaptic testing of the device. (**e**) Endurance test over 500 consecutive reset-set cycles, read at 0.5 V. (**f**) Retention tests for HRS and LRS after reset and set operations, respectively, read at 0.5 V. (**g**) Typical I–V characteristic curves over 50 repeated switching cycles. (**h**) Endurance test over 100 consecutive reset-set cycles, read at 1.1 V. (**i**) Synaptic testing of the device.

3.4. I–V Curve Fitting Analysis

To verify the conduction mechanisms of the devices, we fitted the forward I–V curves of two devices. As shown in Figure 3a, the forward portion of the single-crystal device was fitted according to the space-charge-limited current (SCLC) model. The SCLC model can be represented by the following equation [27]:

$$J = \frac{9\mu\varepsilon_r\varepsilon_0}{8d^3}\left(\frac{\theta_r}{1+\theta_r}\right)V^2 \quad (1)$$

where J is the current density, ε is the dielectric constant, μ is the carrier mobility, and θ is a defect-related constant. In the low voltage region, the slope is approximately 0.88, indicating a sub-linear relationship between current and voltage. This behavior may be related to ohmic conduction or defect state capture currents within the material. In the intermediate voltage region, the current exhibits a nonlinear relationship with increasing voltage, typically associated with the trap-filled limit (TFL) stage. As the voltage increases to a certain level, the trap states begin to fill up, leading to a rapid increase in current. In the higher voltage region, the current transitions to the SCLC region. In this region, the conduction mechanism is governed by the space charge of the injected current, where the relationship between current and voltage typically follows a square or higher power law.

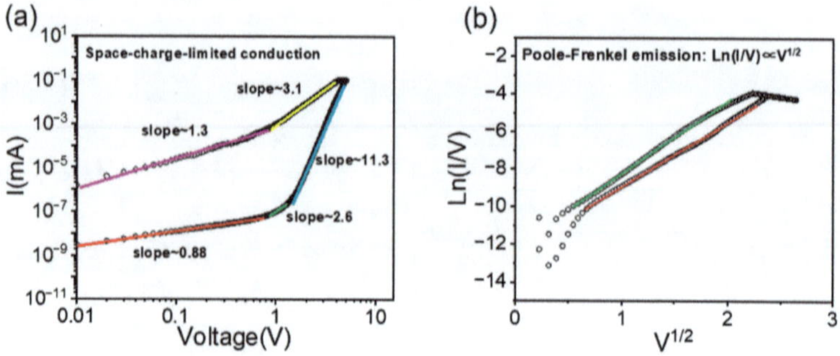

Figure 3. Fitting of I–V curves of the device in the positive voltage range with two different conduction models: (**a**) single-crystal devices: space-charge limited conduction and (**b**) polycrystalline devices: Poole–Frenkel emission.

We fitted the forward I–V curve of the polycrystalline device according to the Poole–Frenkel model, as shown in Figure 3b, which follows the equation [28]:

$$J = AE{\cdot}exp\left(-\frac{B}{kT}{\cdot}E^{\frac{1}{2}}\right),\ ln(I/V) \propto V^{\frac{1}{2}} \quad (2)$$

where J is the current density, E is the electric field intensity, A and B denote constants, and k stands for the Boltzmann constant. The logarithm of the current-to-voltage ratio $ln(I/V)$ exhibits a nearly linear relationship with $V^{1/2}$, indicating the dominance of the Poole–Frenkel mechanism in the conduction process of polycrystalline devices. Based on this model, we hypothesize that defects within the bulk of polycrystalline devices capture electrons and, under the influence of an external field, reduce the energy barrier of the defect states, thereby facilitating the successful excitation of electrons into the conduction band.

Foglietti et al. [29] confirmed the ohmic contact at the CeO_2/NSTO interface and indicated that the redox process occurs at the Pt/CeO_2 interface. Therefore, we believe that the differences in the RS effect between single-crystal and polycrystalline devices are due to the differences in grain boundaries within the SDC film. To verify the above hypothesis, we will also utilize X-ray Photoelectron Spectroscopy (XPS) for validation.

3.5. XPS Analysis and Switching Mechanism Analysis

Using XPS to probe the surface stoichiometric differences of single-crystal and polycrystalline SDC20 thin films, the Ce 3*d* XPS spectra for both cases are depicted in Figure S3.

The Ce 3*d* spectra were fitted using the spin-orbit splitting dipole for $3d_{3/2}$ and $3d_{5/2}$ orbitals [30]. During the fitting process, the intensity ratio of the $3d_{3/2}$ to $3d_{5/2}$ peaks was fixed at 1.5 (with the $3d_{3/2}$ core energy level defined as u and the $3d_{5/2}$ core energy level defined as v), and a spin-orbit splitting binding energy difference Δ = 18.514 eV was employed. Additionally, five pairs of spin-orbit split peaks were fitted. Among these pairs, v_0 and u_0, v' and u', are associated with Ce^{3+}, while v and u, v'' and u'', and v''' and u''' are associated with Ce^{4+}. The fitting data are provided in Table S2. By calculating the ratio of the Ce^{3+} to Ce total peak areas, the Ce^{3+}/Ce_{tol} values for the single-crystal and polycrystalline SDC20 films were determined to be 14% and 12.3%, respectively, as shown in Figure S3a. The fitting results indicate that, at near-surface positions, single-crystalline SDC20 thin films possess a greater abundance of oxygen vacancies compared to polycrystalline SDC20 thin films.

However, in most studies, polycrystalline films typically exhibit a higher concentration of oxygen vacancies compared to single-crystalline films due to the presence of grain boundaries. Therefore, we conducted XPS analysis at various etching depths for both single-crystalline and polycrystalline films and plotted the variation of each element's proportion with their etching depth. As shown in Figure 4a, after etching, a higher concentration of Ce^{3+} is exposed on the surface of the polycrystalline film. By calculating the peak areas, the proportions of each element can be determined. As depicted in Figure 4b, the proportion of Ce^{3+} increases with the etching depth. This increase is attributed to the unavoidable reduction reactions during the etching process and the exposure of Ce^{3+} from the bulk. Similarly, we performed the same characterization on the single-crystalline film. As shown in Figure S3b,c, etching also exposes more Ce^{3+} in the single-crystalline film. However, the proportion of Ce^{3+} stabilizes with continued etching, confirming that the increase in Ce^{3+} in the polycrystalline film during etching is due not only to reduction reactions but also to the exposure of Ce^{3+} from the bulk. At the same etching depth, polycrystalline SDC thin films exhibit a higher concentration of Ce^{3+} compared to single-crystal SDC thin films. This also confirms that polycrystalline SDC thin films have more oxygen vacancies in the bulk phase.

We propose the following explanation: The substitution of high-valence Ce^{4+} ions by low-valence Sm^{3+} ions lead to the generation of many defects, namely oxygen vacancies, to maintain charge neutrality. Under the conditions of 20 mol% doping, the coordination number of Ce is greater than that of Sm, resulting in a greater tendency for oxygen vacancies to be surrounded by Sm^{3+} [31]. Research conducted by Lei et al. has reported that trivalent dopant ions tend to aggregate at grain boundaries [32], thereby leading to a higher concentration of oxygen vacancies near these boundaries, while relatively fewer oxygen vacancies are observed at near-surface positions. Dou et al.'s work demonstrated that conductive channels form near grain boundaries [24]. This finding explains why polycrystalline devices exhibit lower resistance in the high resistance state (HRS) compared to single-crystalline devices (as shown in Figure 2e,h), corroborating the fact that oxygen vacancies accumulate near grain boundaries.

In summary, the speculation that the conduction mechanism of polycrystalline SDC20 devices conforms to the Poole–Frenkel model appears to be valid. TEM results (Figure 4c) indicate the presence of numerous dislocations within the film, where grain boundaries containing abundant dislocation cores form a common [110] tilt axis, consistent with reported findings [25]. In combination with the TEM results, the accumulation of many oxygen vacancies at grain boundaries (as shown in Figure 4d) captures electrons during the conduction process. Under the influence of an external field, electrons are released from these defects, thereby successfully transitioning to the conduction band (as shown in Figure 4e).

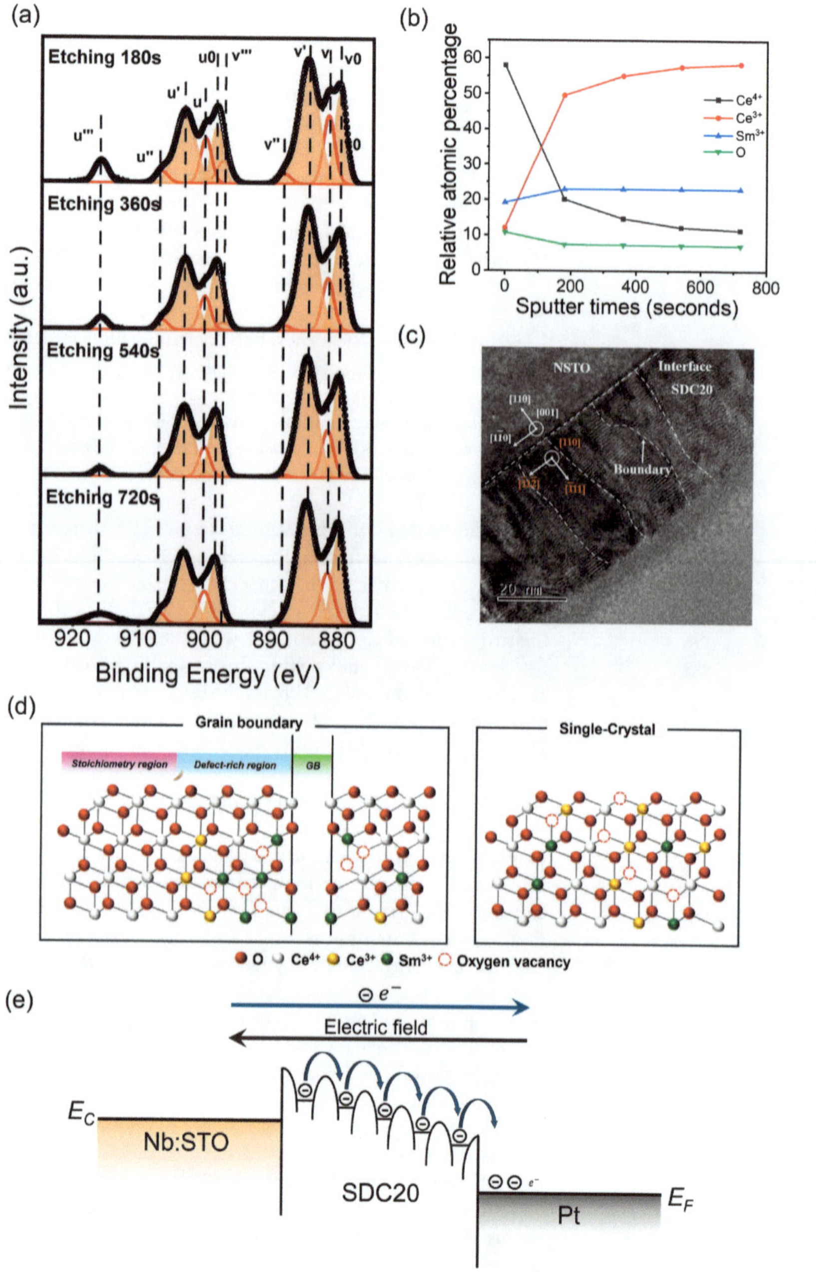

Figure 4. (**a**) The Ce 3*d* XPS spectra of polycrystalline SDC thin films at various etching depths. (**b**) Elemental analysis of polycrystalline SDC thin films at different etching depths. (**c**) TEM images of polycrystalline SDC thin films. (**d**) Schematic illustration of oxygen vacancy distribution within the bulk of SDC20 thin film. (**e**) Conduction mechanism analysis of the polycrystalline device.

4. Conclusions

In summary, we successfully fabricated single-crystalline and polycrystalline SDC films using PLD technology. These films were employed as RS layers to construct Pt/SDC20/NSTO devices, and their RS performance was studied. The single-crystalline device exhibited the best resistive switching performance, maintaining stable resistance states over 500 switching cycles with a switching ratio of 10^2. In contrast, the polycrystalline device demonstrated poorer RS properties, attributed to the accumulation of Sm at grain boundaries, which generated numerous defects that captured electrons, making electron conduction reliant on high-electric fields. This viewpoint was confirmed by XPS and I–V curve fitting. This work highlights the negative impact of grain boundaries in doped systems and provides new insights for the development of advanced memristor devices.

Supplementary Materials: The following supporting information can be downloaded at: https://www.mdpi.com/article/10.3390/ma17133360/s1, Figure S1: (a) XRD patterns of SDC20 powder. (b) The shift of the SDC (220) orientation peak.; Figure S2: SEM images of the single-crystal (a) and polycrystalline (b) SDC20 film. Figure S3: (a) The Ce 3*d* XPS spectra of single-crystal and polycrystalline SDC thin films. (b) Elemental analysis of single-crystal SDC thin films at different etching depths. (c) The Ce 3*d* XPS spectra of polycrystalline SDC thin films at various etching depths. Table S1: Table of ICP results for SDC.; Table S2: Comparison of fitting results of each component of Ce 3*d* XPS spectra of SDC thin films.

Author Contributions: W.S. conducted all experimental analyses and characterizations and prepared the original manuscript. L.W. provided suggestions for XPS analysis. N.Y. provided valuable insights for the mechanism analysis of the devices, as well as for the writing, formatting, and revision of the manuscript. All authors have read and agreed to the published version of the manuscript.

Funding: Funding was obtained via startup funding through ShanghaiTech University and the National Natural Science Foundation of China (52202267).

Institutional Review Board Statement: Not applicable.

Informed Consent Statement: Not applicable.

Data Availability Statement: The raw data supporting the conclusions of this article will be made available by the authors on request.

Acknowledgments: N.Y. acknowledges the startup funding support from ShanghaiTech University. The authors also appreciate the Analytical Instrumentation Center of ShanghaiTech University and CħEM of ShanghaiTech University. The authors are also thankful for funding from the National Natural Science Foundation of China (52202267).

Conflicts of Interest: The authors declare no conflicts of interest.

References

1. Sun, K.; Chen, J.; Yan, X.J.A.F.M. The future of memristors: Materials engineering and neural networks. *Adv. Funct. Mater.* **2021**, *31*, 2006773. [CrossRef]
2. Kim, S.; Kim, H.; Hwang, S.; Kim, M.-H.; Chang, Y.-F.; Park, B.-G. Analog synaptic behavior of a silicon nitride memristor. *ACS Appl. Mater. Interfaces* **2017**, *9*, 40420–40427. [CrossRef]
3. Backus, J. Can programming be liberated from the von Neumann style? A functional style and its algebra of programs. *Commun. ACM* **1978**, *21*, 613–641. [CrossRef]
4. Moore, G.E. Cramming more components onto integrated circuits. *Proc. IEEE* **1998**, *86*, 82–85. [CrossRef]
5. Tseng, T.-Y.; Sze, S.M. *Nonvolatile Memories: Materials, Devices, and Applications, Vol. 2*; American Scientific Publishers: Stevenson Ranch, CA, USA, 2012.
6. Cavallini, M.; Hemmatian, Z.; Riminucci, A.; Prezioso, M.; Morandi, V.; Murgia, M. Regenerable resistive switching in silicon oxide based nanojunctions. *Adv. Mater.* **2012**, *24*, 1197–1201. [CrossRef] [PubMed]
7. Yao, J.; Sun, Z.; Zhong, L.; Natelson, D.; Tour, J.M. Resistive switches and memories from silicon oxide. *Nano Lett.* **2010**, *10*, 4105–4110. [CrossRef]
8. Kim, K.M.; Kim, G.H.; Song, S.J.; Seok, J.Y.; Lee, M.H.; Yoon, J.H.; Hwang, C.S. Electrically configurable electroforming and bipolar resistive switching in Pt/TiO_2/Pt structures. *Nanotechnology* **2010**, *21*, 305203. [CrossRef] [PubMed]
9. Kim, K.M.; Lee, S.R.; Kim, S.; Chang, M.; Hwang, C.S. Self-limited switching in Ta_2O_5/TaO_x memristors exhibiting uniform multilevel changes in resistance. *Adv. Funct. Mater.* **2015**, *25*, 1527–1534. [CrossRef]

10. Rahaman, S.Z.; Lin, Y.-D.; Lee, H.-Y.; Chen, Y.-S.; Chen, P.-S.; Chen, W.-S.; Hsu, C.-H.; Tsai, K.-H.; Tsai, M.-J.; Wang, P.-H. The role of Ti buffer layer thickness on the resistive switching properties of hafnium oxide-based resistive switching memories. *Langmuir* **2017**, *33*, 4654–4665. [CrossRef]
11. Liu, L.; Xiong, W.; Liu, Y.; Chen, K.; Xu, Z.; Zhou, Y.; Han, J.; Ye, C.; Chen, X.; Song, Z. Designing high-performance storage in $HfO_2/BiFeO_3$ memristor for artificial synapse applications. *Adv. Electron. Mater.* **2020**, *6*, 1901012. [CrossRef]
12. Nili, H.; Walia, S.; Balendhran, S.; Strukov, D.B.; Bhaskaran, M.; Sriram, S. Nanoscale resistive switching in amorphous perovskite oxide (a-$SrTiO_3$) memristors. *Adv. Funct. Mater.* **2014**, *24*, 6741–6750. [CrossRef]
13. Campbell, C.T.; Peden, C.H. Oxygen vacancies and catalysis on ceria surfaces. *Science* **2005**, *309*, 713–714. [CrossRef] [PubMed]
14. Han, W.-Q.; Wu, L.; Zhu, Y. Formation and oxidation state of CeO_{2-x} nanotubes. *J. Am. Chem. Soc.* **2005**, *127*, 12814–12815. [CrossRef] [PubMed]
15. Chiu, F.-C.; Lai, C.-M. Optical and electrical characterizations of cerium oxide thin films. *J. Phys. D Appl. Phys.* **2010**, *43*, 075104. [CrossRef]
16. Schmitt, R.; Nenning, A.; Kraynis, O.; Korobko, R.; Frenkel, A.I.; Lubomirsky, I.; Haile, S.M.; Rupp, J.L. A review of defect structure and chemistry in ceria and its solid solutions. *Chem. Soc. Rev.* **2020**, *49*, 554–592. [CrossRef] [PubMed]
17. Di Martino, G.; Demetriadou, A.; Li, W.; Kos, D.; Zhu, B.; Wang, X.; de Nijs, B.; Wang, H.; MacManus-Driscoll, J.; Baumberg, J.J. Real-time in situ optical tracking of oxygen vacancy migration in memristors. *Nat. Electron.* **2020**, *3*, 687–693. [CrossRef]
18. She, Y.; Wang, F.; Zhao, X.; Zhang, Z.; Li, C.; Pan, H.; Hu, K.; Song, Z.; Zhang, K. Oxygen vacancy-dependent synaptic dynamic behavior of TiO_x-based transparent memristor. *IEEE Trans. Electron Devices* **2021**, *68*, 1950–1955. [CrossRef]
19. Hwang, H.-G.; Pyo, Y.; Woo, J.-U.; Kim, I.-S.; Kim, S.-W.; Kim, D.-S.; Kim, B.; Jeong, J.; Nahm, S. Engineering synaptic plasticity through the control of oxygen vacancy concentration for the improvement of learning accuracy in a Ta_2O_5 memristor. *J. Alloys Compd.* **2022**, *902*, 163764. [CrossRef]
20. Rana, A.M.; Ismail, M.; Ahmed, E.; Talib, I.; Khan, T.; Hussain, M.; Nadeem, M. Thickness effect on the bipolar switching mechanism for nonvolatile resistive memory devices based on CeO_2 thin films. *Mater. Sci. Semicond. Process.* **2015**, *39*, 211–216. [CrossRef]
21. Schmitt, R.; Spring, J.; Korobko, R.; Rupp, J.L.M. Design of Oxygen Vacancy Configuration for Memristive Systems. *ACS Nano* **2017**, *11*, 8881–8891. [CrossRef]
22. Lupetin, P.; Giannici, F.; Gregori, G.; Martorana, A.; Maier, J. Effects of grain boundary decoration on the electrical conduction of nanocrystalline CeO_2. *J. Electrochem. Soc.* **2012**, *159*, B417. [CrossRef]
23. Kumar, V.; Chen, W.-F.; Zhang, X.; Jiang, Y.; Koshy, P.; Sorrell, C.C. Properties and performance of photocatalytic CeO_2, TiO_2, and CeO_2–TiO_2 layered thin films. *Ceram. Int.* **2019**, *45*, 22085–22094. [CrossRef]
24. Dou, H.; Hellenbrand, M.; Xiao, M.; Hu, Z.; Kunwar, S.; Chen, A.; MacManus-Driscoll, J.L.; Jia, Q.; Wang, H. Engineering of Grain Boundaries in CeO_2 Enabling Tailorable Resistive Switching Properties. *Adv. Electron. Mater.* **2023**, *9*, 2201186. [CrossRef]
25. Wang, L.; Li, X.; Hu, X.; Chen, S.; Qiu, Z.; Wang, Y.; Zhang, H.; Yu, Y.; Yang, B.; Yang, Y.; et al. Grain boundary boosting the thermal stability of Pt/CeO_2 thin films. *Nano Res.* **2023**, *16*, 3278–3286. [CrossRef]
26. Cavallini, M.; Gentili, D. Atomic vacancies in transition metal dichalcogenides: Properties, fabrication, and limits. *ChemPlusChem* **2022**, *87*, e202100562. [CrossRef] [PubMed]
27. Dongale, T.; Patil, K.; Gaikwad, P.; Kamat, R. Investigating conduction mechanism and frequency dependency of nanostructured memristor device. *Mater. Sci. Semicond. Process.* **2015**, *38*, 228–233. [CrossRef]
28. Jeong, D.S.; Hwang, C.S. Tunneling-assisted Poole-Frenkel conduction mechanism in HfO_2 thin films. *J. Appl. Phys.* **2005**, *98*, 113701. [CrossRef]
29. Foglietti, V.; Yang, N.; Aruta, C.; Di Pietrantonio, F.; Cannatà, D.; Benetti, M.; Balestrino, G. High plasticity reversible resistive switching in heteroepitaxial metal/CeO_{2-x}/Nb: $SrTiO_3$/Ti/Pt structures. *Nanotechnology* **2016**, *27*, 375705. [CrossRef] [PubMed]
30. Burroughs, P.; Hamnett, A.; Orchard, A.F.; Thornton, G. Dalton Transactions, Satellite structure in the X-ray photoelectron spectra of some binary and mixed oxides of lanthanum and cerium. *J. Chem. Soc. Dalton Trans.* **1976**, *17*, 1686–1698. [CrossRef]
31. Sanghavi, R.; Devanathan, R.; Nandasiri, M.I.; Kuchibhatla, S.; Kovarik, L.; Thevuthasan, S.; Prasad, S. Integrated experimental and modeling study of the ionic conductivity of samaria-doped ceria thin films. *Solid State Ion.* **2011**, *204*, 13–19. [CrossRef]
32. Lei, Y.; Ito, Y.; Browning, N.D.; Mazanec, T.J. Segregation effects at grain boundaries in fluorite-structured ceramics. *J. Am. Ceram. Soc.* **2002**, *85*, 2359–2363. [CrossRef]

materials

MDPI

Article

X-ray Characterizations of Exfoliated MoS_2 Produced by Microwave-Assisted Liquid-Phase Exfoliation

Sebastiano Vasi [1,*], Salvatore Vincenzo Giofrè [2], Siglinda Perathoner [2], Domenico Mallamace [2], Salvatore Abate [2] and Ulderico Wanderlingh [1]

1 Department of Mathematical and Computational Science, Physical Science and Earth Science, University of Messina, Viale F. Stagno D'Alcontres 31, I-98166 Messina, Italy; ulderico.wanderlingh@unime.it

2 Department of Chemical, Biological, Pharmaceutical and Environmental Sciences, University of Messina, Viale Ferdinando Stagno D'Alcontres 31, I-98166 Messina, Italy; salvatorevincenzo.giofre@unime.it (S.V.G.); siglinda.perathon@unime.it (S.P.); d.mallamace@unime.it (D.M.); salvatore.abate@unime.it (S.A.)

* Correspondence: sebastiano.vasi@unime.it

Abstract: An X-ray analysis of exfoliated MoS_2, produced by means of microwave-assisted liquid-phase exfoliation (LPE) from bulk powder in 1-methyl-2-pyrrolidone (NMP) or acetonitrile (ACN) + 1-methyl -2-pyrrolidone (NMP) solvents, has revealed distinct structural differences between the bulk powder and the microwave-exfoliated samples. Specifically, we performed X-ray diffraction (XRD) and X-ray photoelectron spectroscopy (XPS) measurements to identify the elements of our exfoliated sample deposited on a Si substrate by drop-casting, as well as their chemical state and its structural crystalline phase. In the exfoliated sample, the peaks pattern only partially resremble the theoretical Miller indices for MoS_2. In contrast, the bulk powder's spectrum shows the characteristic peaks of the 2H polytype of MoS_2, but with some broadening. Notable is the retention of partial crystallinity in the post-exfoliation phases, specifically in the normal-to-plane orientation, thus demonstrating the effectiveness of microwave-assisted techniques in producing 2D MoS_2 and attaining desirable properties for the material. XPS measurements confirm the success of the exfoliation procedure and that the exfoliated sample retains its original structure. The exfoliation process has been optimized to maintain the structural integrity of MoS_2 while enhancing its surface area and electrochemical performance, thereby making it a promising material for advanced electronic and optoelectronic applications ranging from energy storage to sensing devices under ambient conditions.

Keywords: MoS_2; molybdenum disulfide; liquid-phase exfoliation; microwave; XRD; XPS; 2D

Citation: Vasi, S.; Giofrè, S.V.; Perathoner, S.; Mallamace, D.; Abate, S.; Wanderlingh, U. X-ray Characterizations of Exfoliated MoS_2 Produced by Microwave-Assisted Liquid-Phase Exfoliation. *Materials* **2024**, *17*, 3887. https://doi.org/10.3390/ma17163887

Academic Editor: George Kioseoglou

Received: 12 July 2024
Revised: 30 July 2024
Accepted: 31 July 2024
Published: 6 August 2024

1. Introduction

Transition metal dichalcogenides (TMDs) are very promising layered materials that have been widely studied for their scalability and thickness-dependent properties [1–9]. TMDs are compounds with the chemical formula MX_2, where M stands for a transition metal element (Mo, W, Nb) and X indicates a chalcogen (e.g., Se, S, Te). Bulk TMDs have a layered crystal structure assembled by stacking together two-dimensional (2D) X-M-X sheets with weak van der Waals (vdW) forces acting between adjacent layers that can be exfoliated/synthesized to convert layered crystals into 2D nanosheets by means of several techniques and methods [8–26]. In particular, liquid-phase exfoliation (LPE) has been proven to be a very reliable method to obtain nanosheets in liquid media from bulk via intercalation into interlayers, weakening the interlayer bonding, especially if combined with microwave radiation [16,17,27,28]. Among the TMD materials, molybdenum disulfide (MoS_2), which is found abundantly in nature in the form of molybdenite, is one of the most studied [5,19,29,30]. Its crystal structure consists of an hexagonal plane of Mo atoms sandwiched by hexagonal planes of S atoms. These triple planes—a monolayer of MoS_2—stack on top of each other with strong covalent bonds between the Mo and S atoms and weak van der Waals forces holding layers together, until the bulk structure is formed. Bulk

MoS_2 is reported to have an indirect bandgap of 1.2 eV— similarly to silicon—whereas 2D monolayer MoS_2 nanosheets have a direct bandgap of 1.8 eV, possessing intriguing optical properties suitable for several applications, such as light-emitting diodes, transistors, batteries, and photovoltaics [4,10,18,22,27,28,31–42]. In fact, studies of thin and ultrathin films produced with various techniques, starting from exfoliated two-dimensional materials or bulk materials, to create devices and heterostructures that can be used in fields ranging from the medical to the energetic areas are increasingly widespread and frequent [9,12,14,22,23,43–46]. In particular, the study and the realization of third-generation photovoltaic cells that exploit the properties of 2D materials are spreading more and more. For example, 2D vdW materials have been employed to improve utilizing photoexcited hot carriers (HCs)—high-energy carriers beyond the band edges of the semiconductor material used, which are not in thermal equilibrium with the lattice—for light-harvesting applications such as hot carrier solar cells (HCSCs), enhancing the solar to electric power conversion efficiency (PCE) [47–51].

In this frame, a 2D MoS_2 semiconductor material is considered very promising for its excellent optical absorption property and fast charge carrier dynamics, and, consequently, it seems natural how important it is to define solid crystal structures distinguishing between the crystalline phases of our bulk and exfoliated material.

Specifically, bulk molybdenum disulfide crystals are characterized by three different polytypes, 1T, 2H, and 3R phases, where the number indicates how many monolayers are considered internally to the unit cell of the crystal lattice, and the letters T, H, and R refer to the trigonal, hexagonal, and rhombohedral structural symmetry, respectively [2,52]. The 2H phase is shown in Figure 1 as a ball-and-stick representation in which Mo and S atoms are reported in grey and yellow, respectively. Structural phases depend on the stacking arrangement of layers in MoS_2 with respect to Mo coordination. Concerning 2D MoS_2, five polymorphs—1H, 1T, etc.—have been suggested for the atomic structure of monolayers [2,10,52–54]. Then, while the 2H configuration possesses an atomic stacking sequence, a 1H arrangement denotes a single-layer structure of a 2H phase [2,9,54,55]. The structure of 2H-MoS_2 facilitates having several exposed surfaces, one along the edges of SxMoxS layer via the c-axis named the (002) basal plane, while the latter is along the edges of SxMo and is known as the (100) plane [54]. Thus, efficiently distinguishing the presence of a monolayer or multiple monolayers and their respective crystal structure can greatly affect the creation of a device and its characteristics. The most used technique to determine the different phases of bulk and exfoliated materials is X-ray diffraction (XRD) analysis [56].

In this frame, herein we propose a study regarding the X-ray characterizations of exfoliated MoS_2 produced by means of microwave-assisted LPE for the future realization of devices (e.g., HCSC). LPE is fairly well studied, but still there are not many papers in which this technique is assisted by microwaves, especially to obtain two-dimensional materials. Compared with what is present in the literature, in this study, we propose not only variations in the preparation of samples to be exfoliated and their subsequent filtration, but also an extensive study on exfoliated sample characterization by X-ray. In fact, by means of XRD, we obtain important information to distinguish structural differences between the bulk sample to be exfoliated and the exfoliated samples. Furthermore, we support our results obtained with XRD by accompanying them with X-ray photoelectron spectroscopy (XPS) measurements. This technique allows the identification of the elements that exist within a material (elemental composition) or are covering its surface, as well as their chemical state, and the overall electronic structure and density of the electronic states in the material. In our case, our scope is to confirm the success of the exfoliation procedure and that the exfoliated sample retains its original structure. In particular, as shown in Figure 1, we will focus our attention on the 2H phase of MoS_2, investigating the correspondence of the family of (parallel) lattice planes along the c axis between the bulk and exfoliated samples.

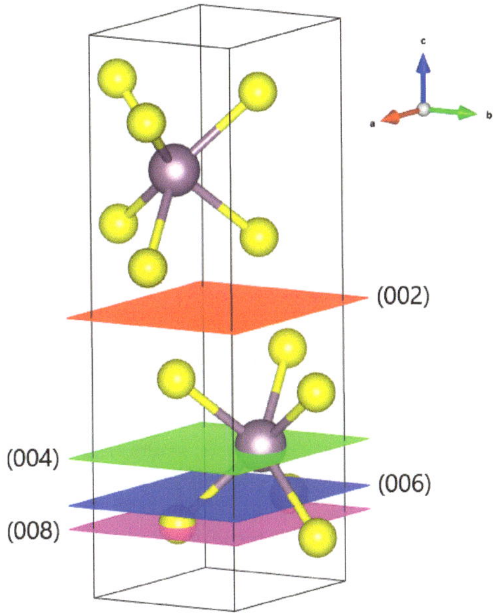

Figure 1. Ball-and-stick schematic model of the 2H phase unit cell for MoS_2 reporting the parallel lattice planes and the corresponding Miller indices along the c axis. This figure is achieved by means of VESTA (Visualization for Electronic and STructural Analysis), a 3D visualization software package for structural models, volumetric data such as electron/nuclear densities, and crystal morphologies [57].

2. Materials and Methods

In our experiments, we made use of model bulk MoS_2 powder (Molybdenum (IV) sulfide, Alfa Aesar) and exfoliated samples obtained by means of microwave-assisted liquid-phase exfoliation (LPE) using a single-mode microwave (MW) synthesis system (Discover from CEM Corp., Matthews, NC, USA).

2.1. Preparation of Exfoliated MoS_2

A microwave-heating-based exfoliation process in the presence of intercalated solvents was used for exfoliating MoS_2. This method is based on the use of a solvent or solvents which intercalate into the interlayer spaces of the nanomaterial and reduce the attractive van der Waals interactions between the layers [15]. A microwave-assisted approach to exfoliate MoS_2 was performed using two methods reported in the literature which involve a pre-treatment of the powder with a wet-grinding procedure using high affinity solvents to avoid powder agglomeration to achieve uniform suspensions [16,17].

2.1.1. Exfoliation of MoS_2 with NMP Solvent (MoS_2-NMP)

The first exfoliation method used was carried out using 1-Methyl-2-pyrrolidinone (NMP), both as a pre-wetting solvent and as a solvent capable of dispersing the exfoliated MoS_2, and then the sealed vessel containing the powder and fluid media was placed into a single-mode microwave reactor and irradiated at 140 W/60 min. After microwave irradiation, the exfoliated sample of MoS_2-NMP was obtained by diluting the suspension with Dimethyl sulfoxide (DMSO) and filtering it with a 0.2 μm polytetrafluoroethylene (PTFE) filter membrane on a sand core filtration system. Then, the resulting residue on the filter was redispersed in ethanol, centrifuged at 5000 rpm/25 min, and the collected supernatant was evaporated under vacuum. The detailed procedure followed to produce the sample MoS_2-NMP is reported below:

- Pre-treatment: An amount of 10.2 mg of bulk MoS_2 powder and 0.2 mL of NMP were first sonicated for 5 min and then transferred into a 10 mL quartz vessel with another 2.8 mL of solvent.
- Microwave exfoliation: MW irradiation with the CEM Discover system and high-speed stirring to homogenize the whole reaction were used. The 'Power Max' function was used at 140 W for 60 min (as operation parameters) with a max temperature of 200 °C, and during microwave irradiation, the temperature was maintained at approximately 100 °C through an air-cooling system.
- Post-Treatment: After cooling, the obtained suspension was first diluted with 5 mL of DMSO, and then filtered with a 0.2 μm PTFE filter membrane on a sand core filtration system; subsequently, ethanol was used to rinse the dust deposited on the PTFE filter. The solid residue was redispersed in ethanol by 5 min of ultrasonic agitation and centrifuged with 5000 rpm/25 min to remove the thick sheets. The supernatant was then dried under vacuum at 40 °C.

2.1.2. Exfoliation of MoS_2 with ACN and NMP Solvents (MoS_2-ACN-NMP)

The MoS_2-ACN-NMP sample was obtained by the microwave irradiation of MoS_2 powder pre-wetted with ACN and dispersed in NMP. Microwave irradiation was performed using an open vessel option by inserting the flask equipped with a refrigerant directly into the reaction cavity at 300 W/12 min. The green-dark suspension was centrifuged with 2000 rpm/20 min, the supernatant was filtered with a 0.2 μm PTFE filter membrane on a sand core filtration system, and an ACN/ethanol mixture was used to rinse the dust deposited on the PTFE filter. The resulting filtered solution was collected and evaporated under reduced pressure at 40 °C to obtain MoS_2-ACN-NMP sample. In the following passages, details of the exfoliation method are given:

- Pre-treatment: An amount of 30.2 mg of bulk MoS_2 powder and 0.5 mL of ACN were sonicated in a glass vial for about 10 min at 30 °C (until most of the solvent had evaporated).
- Microwave exfoliation: The residue pre-wetted with ACN was transferred to a 100 mL round button flask and 30 mL of MNP was added. Microwave irradiation with a single-mode microwave synthesis system and high-speed stirring was performed using an open vessel option by inserting the flask equipped with refrigerant directly into the reaction cavity of the Discover CEM microwave. The 'Power Max' function was used at 300 W for 12 min (as operation parameters) with a max temperature of 150 °C.
- Post-Treatment: After cooling, the green-dark obtained suspension was centrifuged with 2000 rpm/20 min to remove the thick sheets. The supernatant was then filtered with a 0.2 μm PTFE filter membrane on a sand core filtration system, and an ACN/ethanol mixture was used to rinse the dust deposited on the PTFE filter. The resulting filter solution was collected and evaporated under reduced pressure at 40 °C.

2.2. *Morphology Characterization*

The scanning electron microscopy (SEM) image reported in Figure 2 shows an example of our exfoliated sample. The measurement was performed with a Phenom Prox (PhenomWorld)—an all-in-one desktop SEM for imaging and analysis with a fully integrated energy-dispersive spectrometer (EDS)— by using a magnification of 1000× and a working distance (WD) of 15.2 mm. As shown in the figure, we obtained lateral dimensions of almost 17 μm. Furthermore, by EDS measurements, we verified the stoichiometry of our sample, confirming the presence of sulfur and molybdenum.

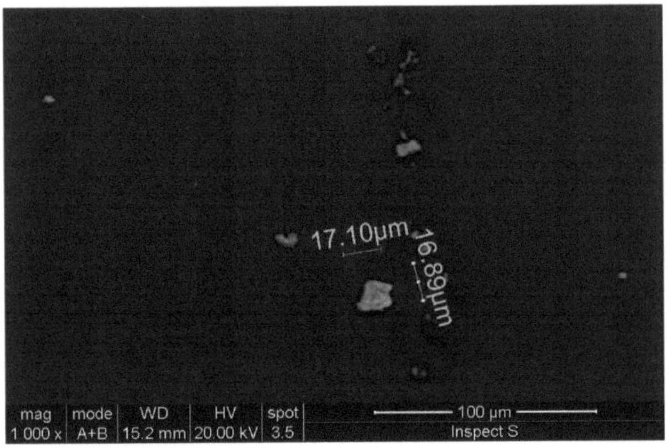

Figure 2. SEM acquisition of our exfoliated sample, showing lateral dimensions of almost 17 µm.

2.3. XRD Analysis

The prepared samples were characterized by X-ray diffraction (XRD) to determine any modifications in the crystal structure of the exfoliated material, with a particular focus on 2H phase. The samples, in the final formulation, consisted of the dispersion of 1.6 mg/mL of exfoliated material in a [1:1] water/ethanol solution. After the sample was briefly vortexed, about 200 µL of such a solution was spread on a [100] Si wafer substrate by the drop-cast method, and used for the measure after complete drying.

The XRD measurements were collected on a Bruker D2 Phaser diffractometer, a bench-top instrument working in Bragg–Bentano geometry with an accuracy of about 0.02° throughout the entire measuring range. This kind of instrument is usually used for routine phase identification and quantitative phase analysis. The X-ray source is a Cu standard sealed ceramic tube with $K\alpha$ at 1.5400 Å($K\alpha_1$) and 1.54439 Å($K\alpha_2$), filtered by a 0.5 mm Ni slab and collimated by a 0.25 mm Soller slit. The diffracted beam was recorded by a Lynxeye detector in the 1D mode, after collimation with a 0.6 mm slit and a knife-edge of 1 mm. The data were recorded from 10° to 75° with 3 s of integration time giving a total run time of about three hours. In this configuration, the Bremsstrahlung contribution is totally removed, but the $K\alpha_2$ is only partially filtered. As a consequence beyond an obvious intensity increment, the presence of two close wavelengths helps to identify faint signals that originated from the sample, which will always be present as doublets, considering that in this kind of application, the sample contribution is expected to be very small compared to the background generated by the sample supporting structure. Moreover, we have experienced that removing the $K\alpha_2$ contribution via software routines results in distorted $K\alpha_1$ peaks in the case of signal with low counts.

The acquired data were analyzed and visualized using Gnuplot, a open-source portable command-line driven graphing utility.

2.4. XPS Analysis

PHI Versa Probe II (Physical Electronics), equipped with an Al $K\alpha$ (1486.6 eV) X-ray source, measured the XPS spectra. The survey spectra were recorded with an analyzer energy path of 117 eV, while the C1s, Mo3d, and S2p core levels were measured at 23.5 eV passing energy. The X-ray beam size was 100 microns at 25 W. A charge neutralization procedure was performed by simultaneous irradiation of samples using a low-energy electron beam and an ion beam before measuring the spectra. The XPS peak position was referenced to a Au metal foil (84.0 eV). XPS peaks were analyzed by using the Multipack Data Reduction Software (https://www.phi.com/surface-analysis-equipment/system-software/multipak-data-reduction-software.html, ULVAC-PHI, Inc., Chigasaki, Japan), employing a Shirley background curve. After the analysis, we made use of Gnuplot to plot the data.

3. Results and Discussion

3.1. XRD Measurements

A panoramic view of the collected diffractograms is presented in Figure 3, in which we reported the spectra acquired for the MoS_2 samples investigated (bulk powder and exfoliated) along with the diffraction pattern calculated for bulk 2H-MoS_2 by means of VESTA using the corresponding CIF file. The expected Miller indices estimated in this latter case are shown in an impulse graph as red vertical lines named Bulk (calc); the blue line represents the experimental XRD data for our raw bulk MoS_2 powder, named Bulk (exp); and the green and light-blue lines refer to the XRD spectra of exfoliated MoS_2-NMP and MoS_2-ACN-NMP pectively.

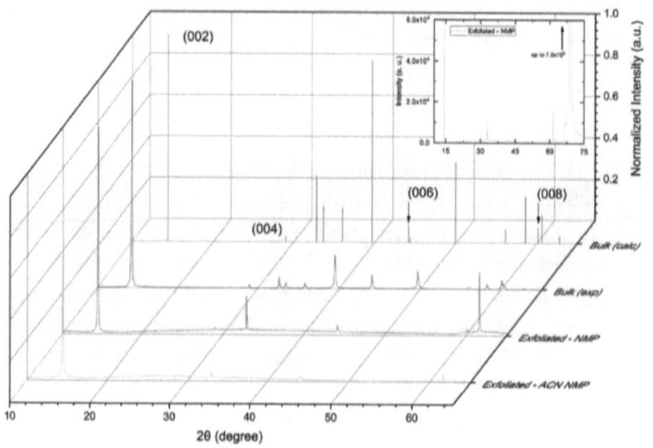

Figure 3. XRD spectra for calculated bulk 2H-MoS_2 [Bulk (calc), red vertical lines], raw bulk MoS_2 powder [Bulk (exp), blue line], MoS_2-NMP [Exfoliated-NMP, green line] and MoS_2-ACN-NMP [Exfoliated-ACN-NMP, light-blue line]. The inset shows the presence of an intense Si peak at about 69°, which is not reported in the 3D graph to keep the normalized peaks of the MoS_2 samples well visible.

As can be observed from the inset of the same figure, the diffractograms are dominated by the [100] Si contribution. That consists of the expected [400] strong peak at 69.1298, generated by $K\alpha_1$ and $K\alpha_2$, extending up to several orders of magnitude. Note that the asymmetry of this peak is due to the sharp rising edge of the absorption of the Ni filter [58]. Close to this peak, there is also the presence of a [400] contamination from the $K\beta$ (1.39220 Å) peak at $2\theta = 61.68966°$. Furthermore there is also the occurrence of the basis-forbidden Si [200] reflection, which can be observed in the range between about 31° and 35°, due to multiple diffraction (Umweganregung) in the classic $\omega - 2\theta$ scan [59].

As for the measured samples, a first glance at Figure 3 confirms that the raw material used is consistent with the expected MoS_2 powder diffraction pattern; moreover, in the exfoliated samples, only the peaks originated by planes parallel to *c* axes are observed, and this suggests that the exfoliation was successful and that the obtained flakes are correctly deposited on the Si surface. For a more detailed analysis of the peak position (i.e., plane spacing) in the exfoliated samples, we relayed on the [400] Si peak position, which always presents as a substrate. This peak has been fitted with the sum of two lorentzians generated by the two $K\alpha$ line reflections by the [400] plane (d = 1.35765Å) at a 2θ of 69.123 and 69.321, respectively. This was performed in order to determine both the instrumental angular shift and the intensity ratio among $K\alpha$ lines. Those values were then used to determine the position of the remaining peaks by means of a fitting procedure that includes both Cu$K\alpha$ lines. The experimental 2θ shift was found to be of the order of $(7\text{–}8 \cdot 10^{-3})°$, which is below instrumental resolution, while for the $K\alpha$ intensity ratio, we obtained 0.46–0.50.

The obtained results are listed in Table 1, in which we report the Miller indexes and the corresponding interplane distances obtained from the literature (CIF data) along with the experimentally found values for the bulk powder sample and the two types of exfoliated samples.

Table 1. Lattice distances in Angstrom.

Miller Index [1]	CIF Data	Powder	NMP	ACN-NMP
[002]	6.1500	6.14348	6.16374	6.15999
[004]	3.0750	3.07439	3.07866	3.07807
[006]	2.0500	2.04923	2.05103	2.05105
[008]	1.5375	1.53672	1.53758	1.53769

[1] Crystal planes refer to the 2H phase typical of the bulk.

Figure 4 reports the fit results for the starting bulk powder sample partly mixed with Si powder (in order to use the [400] Si peak as a reference for accurate calibration). Only the [00n] planes from MoS_2 are considered, since only these are observed after the exfoliation processes. In particular, for the last sub figure, a second component has been added to also include the [112] plane of MoS_2, which falls close to [008]. In Figures 5 and 6, we report the fit results for exfoliated samples using NMP and ACN-NMP as the solvent in the production process. The fitting procedure and the parameters obtained are given in the Supplementary Information. Although the statistical error is generally bigger than in the bulk powder case, the quality of the fits nevertheless allows a reliable assignment of the peak positions and the consequent inter plane distances.

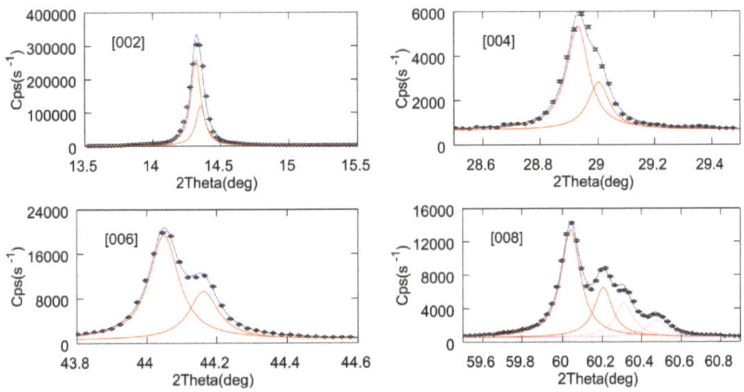

Figure 4. Experimental data and fitted lorentzian doublet for MoS_2 powder.

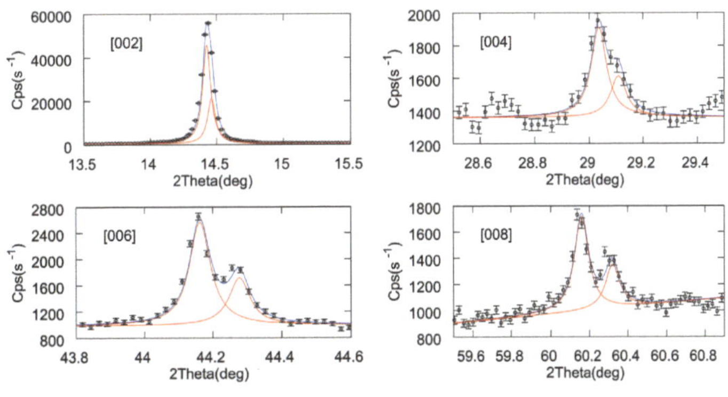

Figure 5. Experimental data and curve fitting of $K\alpha_1$ and $K\alpha_2$ for MoS_2 exfoliated using NMP solvent.

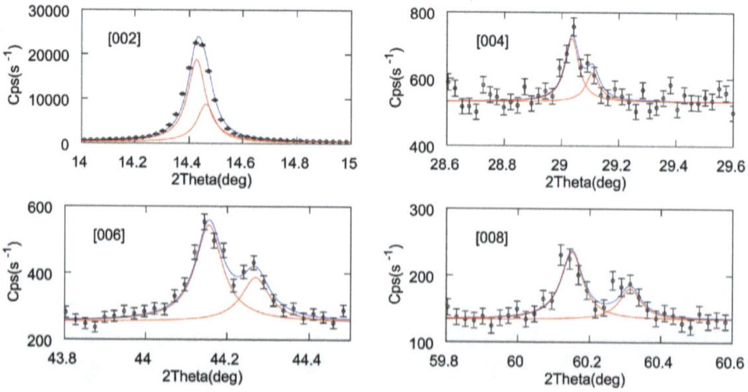

Figure 6. Experimental data and curve fitting of Kα_1 and Kα_2 for MoS_2 exfoliated using ACN-NMP solvent.

3.2. XPS Measurements

The surface characteristics of all the samples were investigated using X-ray photoelectron spectroscopy (XPS). The survey spectra of the bulk MoS_2 (Figure 7d) indicate that the samples primarily contain C, Mo, S, and O elements.

The surface composition in weight percentage together with the atomic ratio S/Mo are reported in Table 2 for both samples investigated. The nearly doubled percentage of carbon and the traces of Na (about 1.80 wt %) observed in the survey of the exfoliated MoS_2-ACN-NMP sample (Figure 7a) are likely related to the solvent used.

Table 2. Surface composition in percentage by weight and atomic ratio S/Mo for powder and exfoliated MoS_2.

Sample	O1s (wt %)	C1s (wt %)	Mo3d (wt %)	S2p (wt %)	Atomic Ratio S/Mo
powder	6.67 ± 0.21	40.02 ± 0.13	31.42 ± 0.18	21.90 ± 0.15	2.04 ± 0.22
exfoliated	16.00 ± 0.22	78.62 ± 0.11	3.18 ± 0.26	2.20 ± 0.27	2.00 ± 0.24

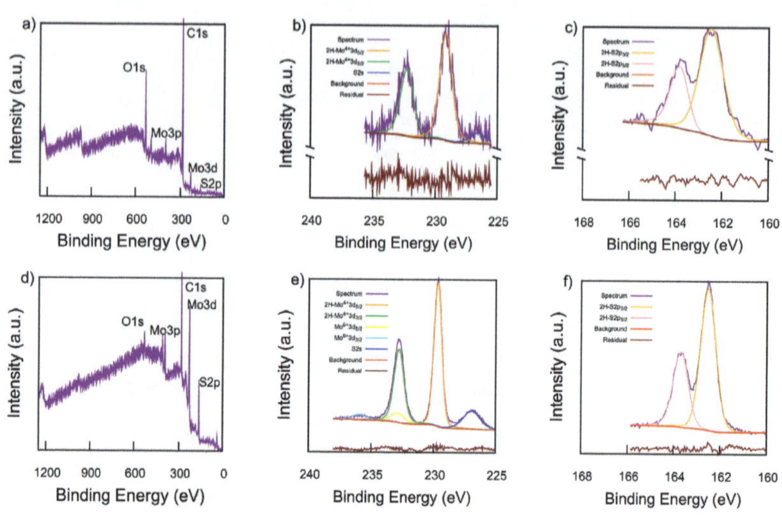

Figure 7. Deconvoluted XPS for exfoliated MoS_2 (**a**) survey, (**b**) Mo3d, (**c**) S2p and bulk MoS_2, (**d**) survey, (**e**) Mo3d, and (**f**) S2p.

The high-resolution Mo3d spectrum of the bulk MoS_2 shown in Figure 7e reveals two main doublet peaks corresponding to the $Mo3d_{5/2}$ and $Mo3d_{3/2}$ states, located at binding energies around 229.67 eV and 232.80 eV, which are consistent with Mo^{4+} species. Additionally, small doublet peaks at 232.85 eV and 235.98 eV indicate the presence of an oxidized Mo^{6+} species. The $Mo3d_{5/2}$-$Mo3d_{3/2}$ splitting energy is 3.13 eV. The narrow full-width-at-half-maximum (FWHM) of 0.73 suggests the presence of the 2H phase (hexagonal crystal structure) according to the literature [60]. The spectrum also displays a peak for S2s at 226.95 eV, which was included in the fitting with the aim to have a good match between fitted curves and raw data, as evidenced by the fitting residual. Similarly, the S2p spectrum in Figure 7f exhibits only the 2H phase with a strong doublet peak at binding energies of 162.50 eV for $S2p_{3/2}$ and 163.69 eV for $S2p_{1/2}$, with a splitting energy of 1.18 eV, typical for metal sulfides [61,62]. The Mo3d spectra of exfoliated MoS_2 (Figure 7b), is slight shifted towards lower binding energies. The deconvoluted spectra reveal the peaks at 229.19 eV for Mo^{4+} $3d_{5/2}$ and 232.32 eV for Mo^{4+} $3d_{3/2}$ [63], confirming the presence of the 2H phase. The modest shift observed in the case of Mo3d can be attributed to the reduction in the oxidation state of Mo, supported by the disappearance of the Mo^{6+} component after the exfoliation treatment [64]. The S2p spectra (Figure 7c) for exfoliated MoS_2 closely resemble those of the bulk sample, displaying a doublet peak at 163.35 eV and 162.15 eV, which correspond to the 2H-$S2p_{1/2}$ and 2H-$S2p_{3/2}$ states, respectively [63]. The Mo/S stoichiometry was also analyzed using XPS data, revealing a sulfur-to-molybdenum ratio of about 2.04 based on the atomic concentration obtained by correlating the respective areas (see Table 2). This stability confirms that the exfoliated sample maintains its original structure, as also reported by other authors [17].

4. Conclusions

The aim of this study was to characterize exfoliated MoS_2 samples, produced by means of microwave-assisted liquid-phase exfoliation (LPE) from bulk powder in 1-methyl-2-pyrrolidone (NMP) or ACetoNitrile (ACN) + 1-methyl-2-pyrrolidone (NMP) solvents, to reveal the correct exfoliation of the sample and demonstrating, in addition, the preservation of the 2H phase after exfoliation with only the retention of the [00n] planes. In fact, as shown in Figure 3, the XRD peaks of our exfoliated samples—compared with the bulk ones—showed a 2H phase and the loss of lattice plans outside the [00n] ones, causing us to believe that only very few layers overlap and the samples are well flattened along the z-axis. Furthermore, as shown in Table 1, which reports the values of the lattice distances obtained by our data analysis with a precise curve fitting procedure, there are only differences in thousandths of Angstrom between all the investigated samples. Furthermore, the XPS analyses on bulk and exfoliated samples support our XRD results, showing the success of the exfoliation procedure and that the exfoliated sample retains its original structure. In this frame, the low-dimensionality and the crystal phase of our exfoliated samples allows us to state that they can be exploited to produce coatings for their use in future energy applications, such as HCSCs.

Supplementary Materials: The following supporting information can be downloaded at: https://www.mdpi.com/article/10.3390/ma17163887/s1.

Author Contributions: Conceptualization, S.V., S.V.G. and U.W.; methodology, S.V.; resources/samples, S.V., S.V.G. and S.P.; investigation/characterization, D.M., S.A. and S.V.; data curation and analysis, S.V., S.A. and U.W.; writing—original draft preparation, S.V. and U.W.; writing—review and editing, S.V., S.V.G., D.M., S.A. and S.P. All authors have read and agreed to the published version of the manuscript.

Funding: This research received no external funding.

Institutional Review Board Statement: Not applicable.

Informed Consent Statement: Not applicable.

Data Availability Statement: The original contributions presented in the study are included in the article, further inquiries can be directed to the corresponding author.

Acknowledgments: This research was conducted within the activities of the RTD-A contract of S. Vasi co-funded by PON 'Ricerca e Innovazione' 2014-2020 (PON R&I FSE-REACT EU), Azione IV.6 'Contratti di ricerca su tematiche Green'.

Conflicts of Interest: The authors declare no conflicts of interest.

Abbreviations

The following abbreviations are used in this manuscript:

TMD	transition metal dichalcogenides
2D	two-dimensional
vdW	van der Waals
LPE	liquid-phase exfoliation
MoS_2	molybdenum disulfide
HC	hot carrier
HCSC	hot carrier solar cell
XRD	X-ray diffraction
XPS	X-ray photoelectron spectroscopy
MW	microwave
NMP	1-methyl-2-pyrrolidone
DMSO	Dimethyl sulfoxide
PTFE	Polytetrafluoroethylene
ACN	Acetonitrile
SEM	scanning electron microscopy
EDS	energy-dispersive spectrometer
VESTA	Visualization for Electronic and STructural Analysis
PTFE	Polytetrafluoroethylene

References

1. Joseph, S.; Mohan, J.; Lakshmy, S.; Thomas, S.; Chakraborty, B.; Thomas, S.; Kalarikkal, N. A review of the synthesis, properties, and applications of 2D transition metal dichalcogenides and their heterostructures. *Mater. Chem. Phys.* **2023**, *297*, 127332. [CrossRef]
2. Zhao, W.; Pan, J.; Fang, Y.; Che, X.; Wang, D.; Bu, K.; Huang, F. Metastable MoS_2: Crystal Structure, Electronic Band Structure, Synthetic Approach and Intriguing Physical Properties. *Chem. A Eur. J.* **2018**, *24*, 15942–15954. [CrossRef] [PubMed]
3. Zhang, W.; Zhang, Y.; Qiu, J.; Zhao, Z.; Liu, N. Topological structures of transition metal dichalcogenides: A review on fabrication, effects, applications, and potential. *InfoMat* **2020**, *3*, 133–154. [CrossRef]
4. Lv, R.; Robinson, J.A.; Schaak, R.E.; Sun, D.; Sun, Y.; Mallouk, T.E.; Terrones, M. Transition Metal Dichalcogenides and Beyond: Synthesis, Properties, and Applications of Single- and Few-Layer Nanosheets. *Accounts Chem. Res.* **2014**, *48*, 56–64. [CrossRef] [PubMed]
5. Ganatra, R.; Zhang, Q. Few-Layer MoS_2: A Promising Layered Semiconductor. *ACS Nano* **2014**, *8*, 4074–4099. [CrossRef] [PubMed]
6. Wang, Q.H.; Kalantar-Zadeh, K.; Kis, A.; Coleman, J.N.; Strano, M.S. Electronics and optoelectronics of two-dimensional transition metal dichalcogenides. *Nat. Nanotechnol.* **2012**, *7*, 699–712. [CrossRef] [PubMed]
7. Niu, T.; Li, A. From two-dimensional materials to heterostructures. *Prog. Surf. Sci.* **2015**, *90*, 21–45. [CrossRef]
8. Manzeli, S.; Ovchinnikov, D.; Pasquier, D.; Yazyev, O.V.; Kis, A. 2D transition metal dichalcogenides. *Nat. Rev. Mater.* **2017**, *2*, 1–15. [CrossRef]
9. Yin, X.; Tang, C.S.; Zheng, Y.; Gao, J.; Wu, J.; Zhang, H.; Chhowalla, M.; Chen, W.; Wee, A.T.S. Recent developments in 2D transition metal dichalcogenides: Phase transition and applications of the (quasi-)metallic phases. *Chem. Soc. Rev.* **2021**, *50*, 10087–10115. [CrossRef]
10. Pariari, D.; Sarma, D.D. Nature and origin of unusual properties in chemically exfoliated 2D MoS_2. *APL Mater.* **2020**, *8*. [CrossRef]
11. Wei, Z.; Li, B.; Xia, C.; Cui, Y.; He, J.; Xia, J.; Li, J. Various Structures of 2D Transition-Metal Dichalcogenides and Their Applications. *Small Methods* **2018**, *2*, 1800094. [CrossRef]
12. Molaei, M.J.; Younas, M.; Rezakazemi, M. Van der Waals heterostructures in ultrathin 2D solar cells: State-of-the-art review. *Mater. Sci. Eng. B* **2022**, *285*, 115936. [CrossRef]
13. Fu, Q.; Han, J.; Wang, X.; Xu, P.; Yao, T.; Zhong, J.; Zhong, W.; Liu, S.; Gao, T.; Zhang, Z.; et al. 2D Transition Metal Dichalcogenides: Design, Modulation, and Challenges in Electrocatalysis. *Adv. Mater.* **2020**, *33*, 1907818. [CrossRef] [PubMed]
14. Velický, M.; Toth, P.S. From two-dimensional materials to their heterostructures: An electrochemist's perspective. *Appl. Mater. Today* **2017**, *8*, 68–103. [CrossRef]

15. Liu, Z.; Wang, Y.; Wang, Z.; Yao, Y.; Dai, J.; Das, S.; Hu, L. Solvo-thermal microwave-powered two-dimensional material exfoliation. *Chem. Commun.* **2016**, *52*, 5757–5760. [CrossRef] [PubMed]
16. Wu, W.; Xu, J.; Tang, X.; Xie, P.; Liu, X.; Xu, J.; Zhou, H.; Zhang, D.; Fan, T. Two-Dimensional Nanosheets by Rapid and Efficient Microwave Exfoliation of Layered Materials. *Chem. Mater.* **2018**, *30*, 5932–5940. [CrossRef]
17. Quirós-Ovies, R.; Laborda, M.; Sabanés, N.M.; Martín-Pérez, L.; Silva, S.M.D.; Burzurí, E.; Sebastian, V.; Pérez, E.M.; Santamaría, J. Microwave-Driven Exfoliation of Bulk 2H-MoS_2 after Acetonitrile Prewetting Produces Large-Area Ultrathin Flakes with Exceptionally High Yield. *ACS Nano* **2023**, *17*, 5984–5993. [CrossRef] [PubMed]
18. Saliba, M.; Atanas, J.P.; Howayek, T.M.; Habchi, R. Molybdenum disulfide, exfoliation methods and applications to photocatalysis: A review. *Nanoscale Adv.* **2023**, *5*, 6787–6803. [CrossRef]
19. Gupta, D.; Chauhan, V.; Kumar, R. A comprehensive review on synthesis and applications of molybdenum disulfide (MoS_2) material: Past and recent developments. *Inorg. Chem. Commun.* **2020**, *121*, 108200. [CrossRef]
20. Mohiuddin, M.; Wang, Y.; Zavabeti, A.; Syed, N.; Datta, R.S.; Ahmed, H.; Daeneke, T.; Russo, S.P.; Rezk, A.R.; Yeo, L.Y.; et al. Liquid Phase Acoustic Wave Exfoliation of Layered MoS_2: Critical Impact of Electric Field in Efficiency. *Chem. Mater.* **2018**, *30*, 5593–5601. [CrossRef]
21. Raza, A.; Hassan, J.Z.; Ikram, M.; Ali, S.; Farooq, U.; Khan, Q.; Maqbool, M. Advances in Liquid-Phase and Intercalation Exfoliations of Transition Metal Dichalcogenides to Produce 2D Framework. *Adv. Mater. Interfaces* **2021**, *8*, 2002205. [CrossRef]
22. Wang, H.; Li, C.; Fang, P.; Zhang, Z.; Zhang, J.Z. Synthesis, properties, and optoelectronic applications of two-dimensional MoS_2 and MoS_2-based heterostructures. *Chem. Soc. Rev.* **2018**, *47*, 6101–6127. [CrossRef] [PubMed]
23. Hoang, A.T.; Qu, K.; Chen, X.; Ahn, J.H. Large-area synthesis of transition metal dichalcogenides via CVD and solution-based approaches and their device applications. *Nanoscale* **2021**, *13*, 615–633. [CrossRef] [PubMed]
24. Wang, J.; Li, T.; Wang, Q.; Wang, W.; Shi, R.; Wang, N.; Amini, A.; Cheng, C. Controlled growth of atomically thin transition metal dichalcogenides via chemical vapor deposition method. *Mater. Today Adv.* **2020**, *8*, 100098. [CrossRef]
25. Sarcan, F.; Fairbairn, N.J.; Zotev, P.; Severs-Millard, T.; Gillard, D.J.; Wang, X.; Conran, B.; Heuken, M.; Erol, A.; Tartakovskii, A.I.; et al. Understanding the impact of heavy ions and tailoring the optical properties of large-area monolayer WS_2 using focused ion beam. *NPJ 2D Mater. Appl.* **2023**, *7*, 23. [CrossRef]
26. Fan, X.; Xu, P.; Zhou, D.; Sun, Y.; Li, Y.C.; Nguyen, M.A.T.; Terrones, M.; Mallouk, T.E. Fast and Efficient Preparation of Exfoliated 2H MoS_2 Nanosheets by Sonication-Assisted Lithium Intercalation and Infrared Laser-Induced 1T to 2H Phase Reversion. *Nano Lett.* **2015**, *15*, 5956–5960. [CrossRef] [PubMed]
27. Sarkar, A.S.; Stratakis, E. Dispersion behaviour of two dimensional monochalcogenides. *J. Colloid Interface Sci.* **2021**, *594*, 334–341. [CrossRef] [PubMed]
28. Chavalekvirat, P.; Hirunpinyopas, W.; Deshsorn, K.; Jitapunkul, K.; Iamprasertkun, P. Liquid Phase Exfoliation of 2D Materials and Its Electrochemical Applications in the Data-Driven Future. *Precis. Chem.* **2024**, *2*. [CrossRef]
29. Nalwa, H.S. A review of molybdenum disulfide (MoS_2) based photodetectors: From ultra-broadband, self-powered to flexible devices. *RSC Adv.* **2020**, *10*, 30529–30602. [CrossRef] [PubMed]
30. Kumar, V.P.; Panda, D.K. Review—Next Generation 2D Material Molybdenum Disulfide (MoS_2): Properties, Applications and Challenges. *ECS J. Solid State Sci. Technol.* **2022**, *11*, 033012. [CrossRef]
31. Ahmed, S.; Yi, J. Two-Dimensional Transition Metal Dichalcogenides and Their Charge Carrier Mobilities in Field-Effect Transistors. *Nano-Micro Lett.* **2017**, *9*, 1–23. [CrossRef] [PubMed]
32. Verduci, R.; Agresti, A.; Romano, V.; D'Angelo, G. Interface Engineering for Perovskite Solar Cells Based on 2D-Materials: A Physics Point of View. *Materials* **2021**, *14*, 5843. [CrossRef]
33. Saha, D.; Kruse, P. Editors' Choice—Review—Conductive Forms of MoS_2 and Their Applications in Energy Storage and Conversion. *J. Electrochem. Soc.* **2020**, *167*, 126517. [CrossRef]
34. Lee, C.H.; Lee, G.H.; van der Zande, A.M.; Chen, W.; Li, Y.; Han, M.; Cui, X.; Arefe, G.; Nuckolls, C.; Heinz, T.F.; et al. Atomically thin p–n junctions with van der Waals heterointerfaces. *Nat. Nanotechnol.* **2014**, *9*, 676–681. [CrossRef]
35. Das, S.; Pandey, D.; Thomas, J.; Roy, T. The Role of Graphene and Other 2D Materials in Solar Photovoltaics. *Adv. Mater.* **2018**, *31*, 1802722. [CrossRef] [PubMed]
36. Bernardi, M.; Palummo, M.; Grossman, J.C. Extraordinary Sunlight Absorption and One Nanometer Thick Photovoltaics Using Two-Dimensional Monolayer Materials. *Nano Lett.* **2013**, *13*, 3664–3670. [CrossRef] [PubMed]
37. Gan, L.Y.; Zhang, Q.; Cheng, Y.; Schwingenschlögl, U. Photovoltaic Heterojunctions of Fullerenes with MoS_2 and WS_2 Monolayers. *J. Phys. Chem. Lett.* **2014**, *5*, 1445–1449. [CrossRef]
38. Matte, H.S.; Gomathi, A.; Manna, A.K.; Late, D.J.; Datta, R.; Pati, S.K.; Rao, C.N. MoS_2 and WS_2 Analogues of Graphene. *Angew. Chem. Int. Ed.* **2010**, *49*, 4059–4062. [CrossRef]
39. Wang, F.; Wang, Z.; Wang, Q.; Wang, F.; Yin, L.; Xu, K.; Huang, Y.; He, J. Synthesis, properties and applications of 2D non-graphene materials. *Nanotechnology* **2015**, *26*, 292001. [CrossRef]
40. Kappera, R.; Voiry, D.; Yalcin, S.E.; Branch, B.; Gupta, G.; Mohite, A.D.; Chhowalla, M. Phase-engineered low-resistance contacts for ultrathin MoS_2 transistors. *Nat. Mater.* **2014**, *13*, 1128–1134. [CrossRef]
41. Splendiani, A.; Sun, L.; Zhang, Y.; Li, T.; Kim, J.; Chim, C.Y.; Galli, G.; Wang, F. Emerging Photoluminescence in Monolayer MoS_2. *Nano Lett.* **2010**, *10*, 1271–1275. [CrossRef] [PubMed]

42. Mak, K.F.; Lee, C.; Hone, J.; Shan, J.; Heinz, T.F. Atomically thin MoS_2: A new direct-gap semiconductor. *Phys. Rev. Lett.* **2010**, *105*, 136805. [CrossRef] [PubMed]
43. Vacik, J.; Ceccio, G.; Lavrentiev, V.; Cannavò, A.; Fink, D.; Kupcik, J.; Svora, P.; Bakardjieva, S.; Plocek, J.; Vasi, S. Surface patterning in thin ternary composites based on Zr, In and C irradiated with 2 MeV W+ ions. *Radiat. Eff. Defects Solids* **2023**, *178*, 94–110. [CrossRef]
44. Massiot, I.; Cattoni, A.; Collin, S. Progress and prospects for ultrathin solar cells. *Nat. Energy* **2020**, *5*, 959–972. [CrossRef]
45. Yang, K.Y.; Nguyen, H.T.; Tsao, Y.M.; Artemkina, S.B.; Fedorov, V.E.; Huang, C.W.; Wang, H.C. Large area MoS_2 thin film growth by direct sulfurization. *Sci. Rep.* **2023**, *13*, 8378. [CrossRef] [PubMed]
46. Sfuncia, G.; Nicotra, G.; Giannazzo, F.; Pécz, B.; Gueorguiev, G.K.; Kakanakova-Georgieva, A. 2D graphitic-like gallium nitride and other structural selectivity in confinement at the graphene/SiC interface. *CrystEngComm* **2023**, *25*, 5810–5817. [CrossRef]
47. Trovatello, C.; Piccinini, G.; Forti, S.; Fabbri, F.; Rossi, A.; De Silvestri, S.; Coletti, C.; Cerullo, G.; Dal Conte, S. Ultrafast hot carrier transfer in WS_2/graphene large area heterostructures. *NPJ 2D Mater. Appl.* **2022**, *6*, 24. [CrossRef]
48. Paul, K.K.; Kim, J.H.; Lee, Y.H. Hot carrier photovoltaics in van der Waals heterostructures. *Nat. Rev. Phys.* **2021**, *3*, 178–192. [CrossRef]
49. Wang, W.; Sui, N.; Chi, X.; Kang, Z.; Zhou, Q.; Li, L.; Zhang, H.; Gao, J.; Wang, Y. Investigation of Hot Carrier Cooling Dynamics in Monolayer MoS_2. *J. Phys. Chem. Lett.* **2021**, *12*, 861–868. [CrossRef]
50. Sulas-Kern, D.B.; Miller, E.M.; Blackburn, J.L. Photoinduced charge transfer in transition metal dichalcogenide heterojunctions—Towards next generation energy technologies. *Energy Environ. Sci.* **2020**, *13*, 2684–2740. [CrossRef]
51. Zhou, H.; Chen, Y.; Zhu, H. Harnessing Hot Carriers in Two-Dimensional Materials. *J. Phys. Chem. C* **2024**, *128*, 9828–9836. [CrossRef]
52. Chhowalla, M.; Shin, H.S.; Eda, G.; Li, L.J.; Loh, K.P.; Zhang, H. The chemistry of two-dimensional layered transition metal dichalcogenide nanosheets. *Nat. Chem.* **2013**, *5*, 263–275. [CrossRef] [PubMed]
53. Pal, B.; Singh, A.; G., S.; Mahale, P.; Kumar, A.; Thirupathaiah, S.; Sezen, H.; Amati, M.; Gregoratti, L.; Waghmare, U.V.; et al. Chemically exfoliated MoS_2 layers: Spectroscopic evidence for the semiconducting nature of the dominant trigonal metastable phase. *Phys. Rev. B* **2017**, *96*, 195426. [CrossRef]
54. Obaida, M.; Hassan, S.A.; Swelam, M.N.; Moussa, I.; Teleb, N.H.; Afify, H.H. Phase manipulation of two-dimensional MoS_2 nanostructures. *Bull. Mater. Sci.* **2023**, *46*, 40. [CrossRef]
55. Voiry, D.; Mohite, A.; Chhowalla, M. Phase engineering of transition metal dichalcogenides. *Chem. Soc. Rev.* **2015**, *44*, 2702–2712. [CrossRef] [PubMed]
56. Shinde, P.V.; Shinde, V.M.; Rout, C.S. X-Ray-Based Techniques for 2D Materials. In *Advanced Analytical Techniques for Characterization of 2D Materials*; AIP Publishing LLCMelville: New York, NY, USA, 2022; pp. 4-1–4-24. [CrossRef]
57. Momma, K.; Izumi, F. VESTA: A three-dimensional visualization system for electronic and structural analysis. *J. Appl. Crystallogr.* **2008**, *41*, 653–658. [CrossRef]
58. Harrington, G.F.; Santiso, J. Back-to-Basics tutorial: X-ray diffraction of thin films. *J. Electroceramics* **2021**, *47*, 141–163. [CrossRef]
59. Zaumseil, P. High-resolution characterization of the forbidden Si 200 and Si 222 reflections. *J. Appl. Crystallogr.* **2015**, *48*, 528–532. [CrossRef] [PubMed]
60. Das, K.; Lohkna, S.; Yang, G.; Ghosh, P.; Roy, S. Sulphur vacancy driven phase conversion of MoS_2 nanosheets for efficient photoreduction of CO_2 under visible light. *J. Mater. Chem. A* **2023**, *11*, 21721–21734. [CrossRef]
61. Ganta, D.; Sinha, S.; Haasch, R.T. 2-D Material Molybdenum Disulfide Analyzed by XPS. *Surf. Sci. Spectra* **2014**, *21*, 19–27. [CrossRef]
62. Wang, X.; Cormier, C.R.; Khosravi, A.; Smyth, C.M.; Shallenberger, J.R.; Addou, R.; Wallace, R.M. In situ exfoliated 2D molybdenum disulfide analyzed by XPS. *Surf. Sci. Spectra* **2020**, *27*. [CrossRef]
63. Lukianov, M.Y.; Rubekina, A.A.; Bondareva, J.V.; Sybachin, A.V.; Diudbin, G.D.; Maslakov, K.I.; Kvashnin, D.G.; Klimova-Korsmik, O.G.; Shirshin, E.A.; Evlashin, S.A. Photoluminescence of Two-Dimensional MoS_2 Nanosheets Produced by Liquid Exfoliation. *Nanomaterials* **2023**, *13*, 1982. [CrossRef] [PubMed]
64. Cheng, X.; Wang, L.; Xie, L.; Sun, C.; Zhao, W.; Liu, X.; Zhuang, Z.; Liu, S.; Zhao, Q. Defect-driven selective oxidation of MoS_2 nanosheets with photothermal effect for Photo-Catalytic hydrogen evolution reaction. *Chem. Eng. J.* **2022**, *439*, 135757. [CrossRef]

materials

MDPI

Article

Structure-Driven Performance Enhancement in Palladium–Graphene Oxide Catalysts for Electrochemical Hydrogen Evolution

Krishnamoorthy Sathiyan *, Ce Gao , Toru Wada , Poulami Mukherjee, Kalaivani Seenivasan and Toshiaki Taniike *

Graduate School of Advanced Science and Technology, Japan Advanced Institute of Science and Technology, 1-1 Asahidai, Nomi 923-1292, Ishikawa, Japan; s2220015@jaist.ac.jp (C.G.); toruwada@jaist.ac.jp (T.W.); poulami@jaist.ac.jp (P.M.); kalaivan@jaist.ac.jp (K.S.)

* Correspondence: sathiyan@jaist.ac.jp (K.S.); taniike@jaist.ac.jp (T.T.)

Abstract: Graphene oxide (GO) has recently gained significant attention in electrocatalysis as a promising electrode material owing to its unique physiochemical properties such as enhanced electron transfers due to a conjugated π-electron system, high surface area, and stable support for loading electroactive species, including metal nanoparticles. However, only a few studies have been directed toward the structural characteristics of GO, elaborating on the roles of oxygen-containing functional groups, the presence of defects, interlayer spacing between the layered structure, and nonuniformity in the carbon skeleton along with their influence on electrochemical performance. In this work, we aim to understand these properties in various GO materials derived from different graphitic sources. Both physiochemical and electrochemical characterization were employed to correlate the above-mentioned features and explore the effect of the location of the palladium nanoparticles (Pd NPs) on various GO supports for the hydrogen evolution reaction (HER). The interaction of the functional groups has a crucial role in the Pd dispersion and its electrochemical performance. Among the different GO samples, Pd supported on GO derived from graphene nanoplate (GNP), Pd/GO-GNP, exhibits superior HER performance; this could be attributed to the optimal balance among particle size, defect density, less in-plane functionalities, and higher electrochemical surface area. This study, thus, helps to identify the optimal conditions that lead to the best performance of Pd-loaded GO, contributing to the design of more effective HER electrocatalysts.

Keywords: graphene oxide; functional groups; Pd nanoparticle; electrocatalysts; hydrogen evolution reaction

Citation: Sathiyan, K.; Gao, C.; Wada, T.; Mukherjee, P.; Seenivasan, K.; Taniike, T. Structure-Driven Performance Enhancement in Palladium–Graphene Oxide Catalysts for Electrochemical Hydrogen Evolution. *Materials* **2024**, *17*, 5296. https://doi.org/10.3390/ma17215296

Academic Editor: Julia A. Baimova

Received: 19 September 2024
Revised: 21 October 2024
Accepted: 29 October 2024
Published: 31 October 2024

1. Introduction

The intellectual catalyst design approach is critical for developing advanced materials and accelerating various catalytic reactions, including those in industrial applications [1]. Among various techniques, particular importance is given to electrochemical water splitting reactions, which create clean, sustainable alternative energy to support the hydrogen economy [2,3]. Water splitting reactions proceed via two half-cell reactions: H^+ ions are reduced at the cathode (hydrogen evolution reaction, HER), and the water is oxidized at the anode (oxygen evolution reaction, OER) [4]. While molecular hydrogen (H_2) is not available naturally, the energy produced is clean and environmentally friendly, gaining interest in fuel cell technology and prompting researchers to necessitate its production on a large scale [5]. Moreover, H_2 generated by the electrochemical approach appears to be greener and more sustainable [6]. Platinum (Pt) is considered the benchmark catalyst for HER in acidic media [7]; however, the limited abundance and high cost restrict its practical applications [8–10]. To reduce pressure on sole Pt, palladium (Pd), having a good affinity for hydrogen, can be an alternative [11,12]. However, the catalytic activity of Pd is lower compared to that of Pt since the Pd–hydrogen system tends to have an internal damage phenomenon due to hydrogen embrittlement that causes microstructural changes and cracking, thus reducing its mechanical integrity [13]. It requires inert support materials that can stabilize the Pd electrodes. In that direction, if the desired performance is achieved, efforts should still be made to reduce the metal content and maximize the HER performance to compensate for the cost to a techno-economic level and facilitate real-time applications. One of the strategies to address this issue is reducing the size of the nanoparticles (NPs) and then further decorating them on a high-surface-area substrate [14].

Graphene oxide (GO), a derivative of graphene, has various oxygen-containing functional groups, such as hydroxyl, epoxy, carbonyl, and carboxylic groups [15]. These functional groups act as binding sites for the reactants and allow transition metal ions to coordinate, making GO an excellent support candidate [16,17]. The utilization of Pd in its NPs form has the superior benefits of minimal usage while maintaining high catalytic efficiency when supported on GO [18]. This, in turn, makes the catalyst synthesis cost-effective. Electrochemical processes like HER benefit from binding Pd NPs on a GO support not only due to the uniform coverage of the metal nanoparticles (M-NPs) but also by improving stability [19,20]. GO supports have special advantages over other carbon supports for dispersing M-NPs. For example, typically existing carboxyl groups at the edges of GO sheets can efficiently serve as nucleation sites for Pd [21]. Hence, Pd NPs at higher concentrations are prone to attach at edge positions. If these groups distribute uniformly on the surface, their interaction can facilitate the anchoring of the Pd NPs, thereby providing a more homogeneous distribution. Likewise, epoxy groups lying on the basal plane of the GO sheet may also act as Pd binding sites [22]. GO supports also allow for controlling the degree of oxidation to maintain a balance between the sp^2 hybridized carbon domains (related to conductivity and gas diffusion) and the functional groups introduced during oxidation (for dispersibility and functionalization). In general, the sp^2 hybridized regions of GO are less functionalized and may not provide strong anchoring sites for Pd NPs, resulting in sparser Pd distribution in these areas. Benefitting from the different functional groups attached, the GO has distinct merits: it is capable of controlling the particle dispersion and morphology—creating more catalytic active sites—and improving the binding between Pd NPs and GO supports for better HER kinetics [23]. However, the actual roles of the functional group, their location (on the edges and/or basal plane), their influence on M-NPs' dispersion, and catalytic performance are insufficiently explored, with more focus given to the M-NPs. Besides functional groups, other structural features like the interlayer spacing between the GO sheets and defects (distortion or vacancy) also demand careful consideration.

This work aims to investigate the physiochemical characteristics of a wide series of Pd-loaded GO catalysts resulting from various graphitic sources and explain their HER performance distinguishing it from our previous work, which focused mainly on the role of reactive functional groups and linker molecules in graphene oxide frameworks (GOF) for stabilizing Pd NPs, specifically in the context of the Suzuki–Miyaura reaction [24]. The present work deals with an electrochemical perspective, whereas the existing literature focuses more on reduced graphene oxide (rGO) rather than GO itself. The impact of functional groups and their influence on electrochemical performance, especially with GO, remains relatively underexplored. The results from this research will evidently show a clearer direction in designing effective HER electrocatalysts based on material characteristics and electrochemical behavior.

2. Materials and Methods

2.1. Materials

Four variations in GO materials were studied in the current research, among which three of them were reported in our previous publication [24]. The sample codes of the GO materials are summarized in Table 1. GO-GNP, GO-graphite45, and GO-graphite150 were prepared from graphite sources with different particle sizes. GO-commercial (dry powder, 50–100 mesh) was purchased from Layer One—Advanced Materials. Graphene nanoplatelets (particle size < 2 μm) were purchased from Strem Chemicals. Graphite (particle size > 45 μm, purity > 98%) was purchased from Wako Pure Chemical Industries. Graphite (particle size >150 μm) was supplied by Sigma Aldrich. The details of the synthetic protocols for GO materials are available in the previous publication [24].

Table 1. Sample codes for GO samples.

Carbon Material	Sample Code
N.A. [a]	GO-commercial
Graphite (>45 μm)	GO-graphite45
Graphene nanoplatelets (<2 μm)	GO-GNP
Graphite (>150 μm)	GO-graphite150

[a] The GO was purchased from a commercial source.

Additionally, a graphene oxide framework (GOF) support was prepared from GO-GNP as a reference sample with the expectation of better dispersion and stabilization of Pd NPs in the confined spaces created by the covalently bonded linker molecules [25]. The GOF sample is referred to as GOF-GNP.

A Pd catalyst was synthesized by reducing $PdCl_2$ in the presence of GO and GOF materials in toluene [25]. In this method, the reactive functional groups present on the support were exploited as reductant. For further details about the protocol, refer to our previous publication [24]. The catalyst where Pd was supported on a GO material is referred to as Pd/GO, and that for GOF-GNP is referred to as Pd@GOF-GNP.

Other reagents including diethyl ether (Wako Pure Chemical Industries, Osaka, Japan), palladium (II) chloride ($PdCl_2$, Sigma Aldrich, St. Louis, MO, USA), phenyl diboronic acid (purity > 95.0%, Sigma Aldrich), hydrogen peroxide (35% aqueous solution, Tokyo Chemical Industry Co., Ltd., Tokyo, Japan), sulfuric acid (H_2SO_4) (Kanto Chemical Co., Inc., Tokyo, Japan), potassium permanganate (Kanto Chemical Co., Inc.), toluene (Kanto Chemical Co., Inc.), methanol (Kanto Chemical Co., Inc.), ethanol (Kanto Chemical Co., Inc.), propanol (Fujifilm Wako Pure Chemical Co., Osaka, Japan), 5% Nafion ^{TM} dispersion solution DE520 CS-type (Fujifilm Wako Pure Chemical Co.), and 0.3 μm and 0.05 μm alumina (Al_2O_3-Baikowski Co., Inc., Malakoff, TX, USA) were used without further purification.

2.2. Characterization

The FT-IR spectra were acquired using JASCO 6100 (JASCO, Tokyo, Japan) in the range of 4000–400 cm^{-1} with a resolution of 4 cm^{-1} by 16 scans. The sample powder was ground with dried KBr at a weight ratio of 1:50 and then pressed into a pellet. The Raman spectra of powder samples dispersed on a glass plate were recorded using a laser Raman spectrometer, NRS-4100 (JASCO), with an excitation wavelength of 532 nm and an exposure time of 25 s with 10 acquisitions.

The morphology, particle size, and particle size distribution of the Pd NPs were investigated by TEM using H-7650 (Hitachi, Tokyo, Japan) operating at an acceleration voltage of 100 kV. The samples were dispersed in ethanol with ultrasonication for 15 min, then dropped onto a carbon-coated copper grid and naturally dried overnight.

XPS was used to investigate the chemical composition and the oxidation state of individual elements. The spectra were recorded on Kratos AXIS Ultra DLD (Shimadzu, Kyoto, Japan) equipped with an Al-Kα anode. Powder samples were loaded onto a sample holder using double-sided adhesive copper tape. The survey spectrum was recorded with a 1 eV step size and a pass energy of 80 eV. Narrow scans were recorded with a 0.1 eV step size and a pass energy of 160 eV. The binding energies were calibrated using the C 1s peak of graphitic carbon at 284.6 eV. The sample powder was spread evenly on copper tape and measurement positions were determined by adjusting the stage position to the location where the C 1s peak of the sample exhibited the highest intensity. The spectra were analyzed using XPSpeak 4.1 software, with baseline correction performed using the Shirley method. The elemental composition was calculated by the area ratio of the detected peaks. Note that the Cu peak was not detected in any of the samples, ruling out the possibility that the observed C and oxygen-containing functional group originated from the copper tape.

2.3. Electrode Preparation

Electrochemical measurements were performed with an electrochemical workstation, HZ-7000, from Meiden Hokuto (Tokyo, Japan). The rotation speed was controlled with a rotating electrode device, HR series HR-500, connected to the workstation. A three-electrode cell was used, with a rotating disk electrode (RDE, diameter 5.0 mm and 0.196 cm^2 surface area) as the working electrode, an Ag/AgCl electrode (3.0 M KCl) as the reference, and a graphite rod as the counter electrode. All the experiments were performed at room temperature. Before the electrochemical measurements, the RDE working electrode was polished with 0.3 μm Al_2O_3 to remove any deposited materials, followed by polishing with 0.05 μm Al_2O_3 to make a smooth surface. Then, it was sonicated in water for 30 s using an ultrasonic bath to remove any deposited particles from the surface. A total of 10 mg of catalyst powder was initially dispersed in a mixture containing 1.50 mL of H_2O, Milli-Q water (ultrapure water produced by a Milli-Q purification system, Millipore, Burlington, MA, USA), 480 μL propanol, and 20 μL Nafion ^{TM} binder (5.0 wt%). The solution was ultrasonicated with a bath sonicator for 45 min to attain a uniform dispersion, followed by 5 min of probe sonication. Finally, 10 μL of the prepared ink was drop cast on the RDE surface, covered with a vial, and naturally dried at room temperature. The catalyst loading was set to 0.255 mg cm^{-2}.

2.4. Electrochemical Measurements

Linear sweep voltammetry (LSV) polarization curves were measured using 0.5 M H_2SO_4 as the electrolyte at a scan rate of 5 mV s^{-1} and with a rotation speed of 1600 rpm to study the HER performance of the catalysts. During the measurement, the electrolyte was continuously purged with nitrogen gas using a surface bubbler to prevent atmospheric oxygen. The Tafel slope was obtained from the LSV polarization curve using the logarithmic relationship between overpotential and current density. The double-layer capacitance (C_{dl}) values were calculated by performing cyclic voltammetry (CV) at different scan rates ranging between 10 and 100 mV s^{-1} in a non-Faradaic region and by plotting the $\Delta j = (J_a - J_c)$ vs. scan rate, where J_a and J_c are the current measured at the anodic and cathodic regions, respectively. The electrochemical active surface area (ECSA) was calculated using the following relationship [26],

$$\text{ECSA} = \text{Geometrical Surface area} \times C_{dl}/C_s$$

where C_s is the specific capacitance value (40 μF cm^{-2}) [27] and the geometric surface area is 0.196 cm^2.

During electrochemical measurements, the rotation of the working electrode and the evolved bubbles can cause the coated catalyst to peel off. Therefore, the durability test was performed on a 1×1 cm^2 carbon paper substrate as the working electrode, where the catalyst loading normalized concerning the electrode surface area remains consistent.

The potential specified in this work is referenced to the reversible hydrogen electrode (RHE).

$$\mathbf{E}\text{ (vs. RHE)} = \mathbf{E}\text{ (vs. Ag/AgCl)} + 0.059 \times \text{pH} + 0.197\text{ V}$$

Electrochemical impedance spectroscopy (EIS) measurements were performed with an electrochemical workstation, HZ-Pro S4, from Meiden Hokuto. The measurements were carried out at an open circuit potential in a frequency range from 20 kHz to 0.01 Hz with an AC amplitude of 10 mV. All the polarization curves shown are without i-R compensation.

3. Results and Discussion

3.1. Structural Characterizations

All the GO and Pd/GO samples other than Pd/GO-graphite150 and their support were already characterized in the previous publication [24] so their differences are only briefly explained here. The IR and Raman spectra of the GO samples are shown in Figure 1a,b. The GO samples prepared in-house (GO-GNP, GO-graphite45, and GO-graphite150) had in-plane functional groups such as epoxy (C-O-C) and hydroxyl (C-OH), while GO-commercial possessed a smaller number of in-plane functional groups but many aggregated cyclic esters. GO-GNP was the most defective due to the presence of many edges and GO-commercial was the least defective due to the largest lateral size of the sheets and the least in-plane functionalization.

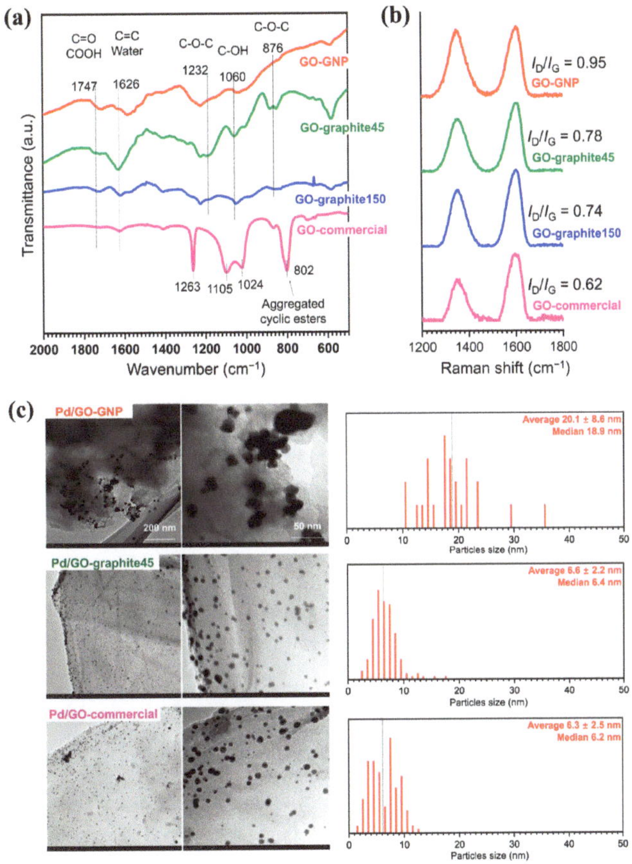

Figure 1. Characterization of the GO supports and the relevant Pd catalysts: (**a**) FT-IR and (**b**) Raman spectra of the supports; (**c**) TEM images for the supported Pd catalysts in two different magnifications. The particle size distribution of Pd NPs is shown on the right side of each image. The particle size was determined by analyzing 200 randomly selected particles at fixed magnification using ImageJ software [28]. The IR and Raman spectra as well as the TEM images for GO-GNP, GO-graphite45, GO-commercial, and the relevant Pd catalysts are from our previous publication [24].

Figure 1c shows the TEM images of the prepared Pd/GO catalysts together with the particle size distribution of Pd NPs calculated from the image. Pd/GO-GNP showed the largest particle size (average diameter = 20.1 $\pm$ 8.6 nm) among the catalyst samples. The reason for the smaller Pd NPs of Pd/GO-graphite45 (6.61 $\pm$ 2.2 nm) and Pd/GO-commercial (6.31 ± 2.5 nm) is probably the formation of Pd NPs between stacked GO sheets. GO-GNP had almost no such space due to the severe exfoliation, which resulted in the formation of significantly coarse Pd NPs.

The XPS elemental analysis results (Table 2) show that GO-GNP was less functionalized compared to the other GO supports as it exhibited ca. 3–4 wt% lower oxygen content. This tendency is maintained even after Pd loading. The higher Pd loading of Pd/GO-GNP and Pd/GO-commercial is attributed to the presence of rough Pd particles deposited on the outer surface [24].

Table 2. Elemental composition determined by XPS.

	C (wt%)	O (wt%)	Cl (wt%)	Pd (wt%)
GO-GNP [a]	65.5	34.5		
GO-graphite45 [a]	61.4	38.6		
GO-graphite150	55.9	44.1		
GO-commercial [a]	62.1	37.9		
Pd/GO-GNP [a]	49.7	25.1	8.6	16.7 (16.0/0.6) [b]
Pd/GO-graphite45 [a]	57.1	29.3	4.0	9.7 (8.9/0.7) [b]
Pd/GO-graphite150	45.8	39.6	3.33	11.3 (7.5/3.8) [b]
Pd/GO-commercial [a]	44.1	30.1	8.4	17.5 (15.7/1.7) [b]

[a] These data are from our previous publication [24]. [b] The content of Pd(0) and Pd(II) are shown in parentheses as (Pd(0)/Pd(II)).

3.2. Electrochemical Performance

The HER performances of Pd/GO catalysts were evaluated using the LSV technique with a 1600 rpm rotation speed. The catalytic current at 10 mA cm^{-2} was fixed as the benchmark for comparing the performance of different catalysts. Among all the catalysts, Pd/GO-GNP showed the best catalytic performance for HER with a lower overpotential (143 mV@10mA cm^{-2}) than the others, Figure 2a. The activity trend followed in the order of Pd/GO-GNP > Pd/GO-graphite45 > Pd/GO-graphite150 > Pd/GO-commercial. Control experiments were performed using GOs without the Pd loading under identical conditions (Figure S1). The resultant LSV curves confirmed that no significant activity was brought by GO alone, suggesting that GO itself is less conductive. The improved activity may arise from the combined benefits of Pd NPs and GO support. It can be anticipated that when supported on GO, Pd/GO catalysts exhibit enhanced catalytic performance because of better Pd dispersion and electronic interactions with GO and by offering high surface-to-volume ratios, with an increased number of catalytic sites available for reactions.

The Tafel slopes obtained from the polarization curves were used to analyze the intrinsic nature of the catalysts. The Pd/GO-GNP had a lower Tafel value (147 mV dec^{-1}) compared to Pd/GO-graphite45 (162 mV dec^{-1}) and Pd/GO-graphite150 (200 mV dec^{-1}), which indicates its faster HER kinetics, Figure 2b. This, in turn, suggests that electron transfer is more facilitated in Pd/GO-GNP than in other catalysts. Note: The Tafel slope specifically measures the rate-determining step's sensitivity to the overpotential and not the overall activity or current density. The Pd/GO commercial catalyst has a low Tafel slope with a lower current density, indicating less overall H_2 production despite favorable reaction kinetics. This may be accounted for by its limited number of exposed active sites. Compared to other forms of GO, GO-GNP generally contains fewer in-plane defects, thereby retaining higher structural integrity, allowing the electron pathways to remain less disrupted and electron transfer to occur more smoothly and efficiently across the material. One possibility is that the interactions between Pd NPs and the GNP surface create an electronically favorable environment, modifying the electronic structure of Pd and thus enhancing its catalytic property to facilitate faster electron transfer. To validate this claim, the EIS technique was used to probe the interfacial electron transfer kinetics of the catalysts. The impedance spectra were represented in a Nyquist plot, Figure S2a, and fitted with a Randles circuit, R(QR), shown schematically in Figure S2b. The elements used for fitting and its summarized results are tabulated in Table S1. The solution resistance (R_s) remains the same for all the catalysts. The charge transfer resistance (R_{ct}) can be easily visualized in Figure S2a. The smaller the semicircle, the smaller the R_{ct}, and the higher the charge transfer. In other words, R_{ct} is inversely proportional to the current density. Pd/GO-GNP exhibited the lowest R_{ct} value, confirming the highest conductivity among the tested catalysts. The results are thus consistent with the Tafel plot. The larger particle size of GO and stacking of GO sheets together hinder the electron transfer process in all the Pd-loaded GO catalysts except for Pd/GO-GNP.

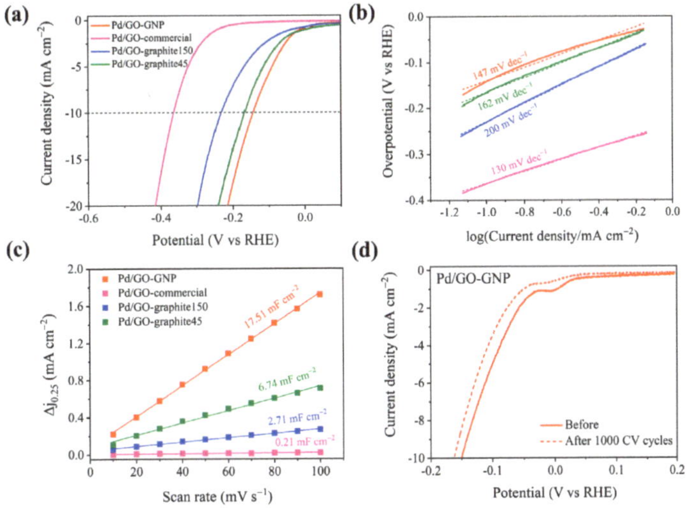

Figure 2. Electrocatalytic activity of synthesized catalysts: (**a**) HER polarization curves measured in 0.5 M H_2SO_4 at 1600 rpm rotation, its corresponding (**b**) Tafel plots, (**c**) C_{dl} plot obtained from the CV (Figure S3), (**d**) durability test: polarization curves obtained before and after 1000 CV cycles at 5 mV s^{-1} scan rate.

The C_{dl} was determined to shed deeper insight into improved performance, giving better knowledge about electrochemically active sites of samples. Figure 2c shows the C_{dl} values obtained by performing CV at various scan rates (Figure S3). Among them, Pd/GO-GNP had the largest C_{dl} value of 17.51 mF cm^{-2} compared with Pd/GO-commercial (0.21 mF cm^{-2}), Pd/GO-graphite45 (6.74 mF cm^{-2}), and Pd/GO-graphite150 (2.71 mF cm^{-2}). Since electrochemical surface area (ECSA) is proportional to the C_{dl} values, the ECSA values were calculated and are displayed in Table S2. In general, the presence of functional groups on GO can facilitate the uniform dispersion and anchoring of Pd NPs with many accessible active sites for catalytic reaction by acting as nucleation sites. It should be highlighted that Pd loading on Pd/GO-GNP, as evidenced by the TEM image, is onto the surface of GO-GNPs, unlike any other type of GO material wherein Pd NPs are stacked between the layers. In addition, the size of the GO-GNP is noteworthy as it is smaller—but higher in surface area—compared to other GOs, making it more accessible for electrochemical reactions. Since the surface topology can influence the HER performance, the obtained LSV results were normalized by the ECSA value to determine the actual source of activity. The obtained ECSA-normalized HER plot (Figure S4) revealed that the HER activity trend is different while normalized with ECSA values. The HER activity followed the order: Pd/GO-commercial > Pd/GO-graphite150 > Pd/GO-GNP > Pd/GO-graphite45. These observations support the fact that the electrochemical activity obtained is mainly due to the enhanced electrochemical surface area of the catalyst, for which one cannot rule out the role of functional groups that indirectly contribute to achieving a high electrochemical surface area.

Summarizing all the electrochemical measurements, it is reasonable to conclude that poorer functionality generates faster charge transfer kinetics; however, to validate this hypothesis, we further expanded our studies to evaluate the electrochemical performance of two types of GO-GNP, i.e., Pd/GNP and Pd@GOF-GNP. The extent of functionalization in GOF is usually higher than in GO due to the introduction of cross-linkers or other functional groups, providing a porous or extended network. As expected from the electrochemical HER results (Figure S5), a better activity of Pd/GO-GNP compared to Pd/GNP and Pd@GOF-GNP is determined. Although Pd@GOF-GNP has well-dispersed Pd NPs, their confinement between the GO layers makes it less accessible for the catalytic reaction. This is further evident from the obtained C_{dl} values displayed in Figure S6. The lower C_{dl} value of Pd@GOF-GNP compared to Pd@GO-GNP could be attributed to its fewer exposed electrochemically active sites. CV measurements (Figure S7) were performed to identify the distinct role of Pd NPs. The palladium peak at +0.1 V vs. RHE reflects that Pd/GO-GNP has more exposed active sites than Pd/GO-graphite45 and other catalysts. This finding correlates well with the TEM results. Altogether, it could be concluded that the Pd NPs' distribution on or in between the GO layers has a critical role in determining HER performance, suggesting that tailoring the functional groups for more exposure to the active surface might show better electrochemical performance.

Stability is an important criterion when proposing an electrocatalyst for practical applications. To analyze it, the HER polarization curves were measured before and after 1000 CV cycles. Figure 2d shows the HER polarization curves of Pd/GO-GNP measured before the 1st CV cycle (solid line) and after (dotted line) the 1000th CV cycle. The result reveals only an increase of 13 mV in overpotential, which implies reasonable stability. Such activity loss could be related to bubble formation that may lead to the catalyst leaching during the HER process. To rule this out, the catalyst with the same catalyst loading was coated on carbon paper, and the stability test was carried out. Figure S8 shows almost no change in the polarization curve, thus confirming excellent durability. The interaction of the Pd NPs with GO's functional groups prevents their sintering or aggregation during catalytic reactions, therefore remaining at a high surface area over time. No characteristic changes were observed in Pd NPs, which were further corroborated by the TEM images shown in Figure 3, displaying Pd/GO-GNP as a better electrocatalyst. XRD analyses were performed before and after the chronoamperometry test, Figure S9a–c. The results reveal that a negative shift is observed in the Pd peaks, suggesting the formation of Pd-hydride species [29,30].

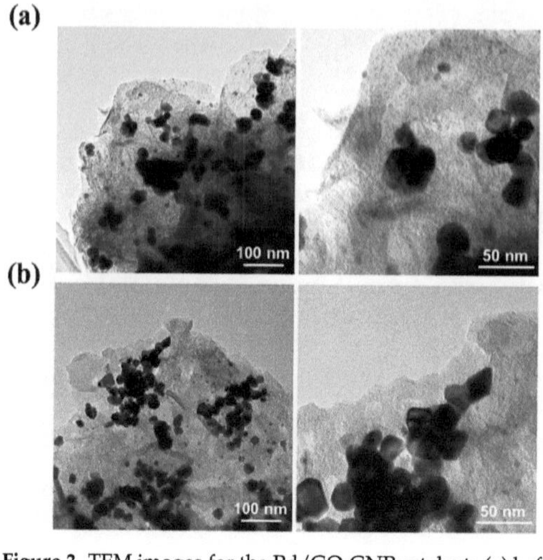

Figure 3. TEM images for the Pd/GO-GNP catalysts (**a**) before and (**b**) after 1000 CV cycles.

4. Conclusions

In summary, we state the critical role of functional groups and other structural features of graphitic-material-supported catalysts—Pd/GO—in electrochemical HER performance. Some key findings of this study are summarized as follows: 1. GO support in-plane functionalities have a significant role in Pd dispersion and anchoring. While the surfaces of Pd NPs are exposed on the lower-functionality GO-GNP support, for other GOs—although Pd NPs are small and homogeneous—the Pd NPs are entrapped between the layers, making them less accessible. 2. Apparently, the lateral dimensions of GO are very important in deciding its catalytic performance. While small GO/GNP sheets form a well-connected network with a high electrochemically active surface area for fast electron transfer, larger sheets in GO-graphite45 and GO-graphite150 hinder this. The Pd/GO-GNP, having the highest activity of 143 mV@10mA cm^{-2}, benefits primarily from the higher electrochemical surface area together with optimal particle size, less in-plane functionalities, and defect density, which ultimately contributes to achieving it. The present work outlines the possibility of using GO supports in the rational design of HER electrocatalysts by tailoring their physiochemical properties.

Supplementary Materials: The following supporting information can be downloaded at: https://www.mdpi.com/article/10.3390/ma17215296/s1. Electrochemical measurements such as LSV, Nyquist plot, C_{dl}, CV measurements, and stability test are given in Figure S1 to S9 along with Table S1, showing the EIS results, and Table S2 displaying the ECSA values derived from C_{dl}.

Author Contributions: K.S. (Krishnamoorthy Sathiyan): Formal analysis, methodology, data curation, investigation, validation, writing—original draft preparation, writing—review and editing, C.G.: methodology, data curation, writing—original draft preparation, T.W.: formal analysis, investigation, writing—review and editing, validation, P.M.: data curation, writing—review and editing, K.S. (Kalaivani Seenivasan): formal analysis, investigation, T.T.; conceptualization, resources, writing—review and editing, project administration, supervision. All authors have read and agreed to the published version of the manuscript.

Funding: No specific external funding sources have been used to conduct this study.

Institutional Review Board Statement: Not applicable.

Informed Consent Statement: Not applicable.

Data Availability Statement: The original contributions presented in the study are included in the article/supplementary material, further inquiries can be directed to the corresponding authors.

Conflicts of Interest: The authors declare no conflicts of interest.

References

1. Prytkova, A.; Kirsanova, M.A.; Kiiamov, A.G.; Tayurskii, D.A.; Dimiev, A.M. Ni–Pd Nanocomposites on Reduced Graphene Oxide Support as Electrocatalysts for Hydrogen Evolution Reactions. *ACS Appl. Nano Mater.* **2023**, *6*, 14902–14909. [CrossRef]
2. Fei, H.; Dong, J.; Arellano-Jiménez, M.J.; Ye, G.; Kim, N.D.; Samuel, E.L.; Peng, Z.; Zhu, Z.; Qin, F.; Bao, J.; et al. Atomic cobalt on nitrogen-doped graphene for hydrogen generation. *Nat. Commun.* **2015**, *6*, 8668. [CrossRef] [PubMed]
3. Li, X.; Zhao, L.; Yu, J.; Liu, X.; Zhang, X.; Liu, H.; Zhou, W. Water Splitting: From Electrode to Green Energy System. *Nano-Micro Lett.* **2020**, *12*, 131. [CrossRef] [PubMed]
4. Jayabharathi, J.; Karthikeyan, B.; Vishnu, B.; Sriram, S. Research on engineered electrocatalysts for efficient water splitting: A comprehensive review. *Phys. Chem. Chem. Phys.* **2023**, *25*, 8992–9019. [CrossRef] [PubMed]
5. Islam, A.; Islam, T.; Mahmud, H.; Raihan, O.; Islam, S.; Marwani, H.M.; Rahman, M.M.; Asiri, A.M.; Hasan, M.; Hasan, N.; et al. Accelerating the green hydrogen revolution: A comprehensive analysis of technological advancements and policy interventions. *Int. J. Hydrogen Energy* **2024**, *67*, 458–486. [CrossRef]
6. Aslam, S.; Rani, S.; Lal, K.; Fatima, M.; Hardwick, T.; Shirinfar, B.; Ahmed, N. Electrochemical hydrogen production: Sustainable hydrogen economy. *Green Chem.* **2023**, *25*, 9543–9573. [CrossRef]
7. Jawhari, A.H.; Hasan, N. Nanocomposite Electrocatalysts for Hydrogen Evolution Reactions (HERs) for Sustainable and Efficient Hydrogen Energy—Future Prospects. *Materials* **2023**, *16*, 3760. [CrossRef]
8. Davodi, F.; Cilpa-Karhu, G.; Sainio, J.; Tavakkoli, M.; Jiang, H.; Mühlhausen, E.; Marzun, G.; Gökce, B.; Laasonen, K.; Kallio, T. Designing of low Pt electrocatalyst through immobilization on metal@C support for efficient hydrogen evolution reaction in acidic media. *J. Electroanal. Chem.* **2021**, *896*, 115076. [CrossRef]

9. Li, Z.; Ge, R.; Su, J.; Chen, L. Recent Progress in Low Pt Content Electrocatalysts for Hydrogen Evolution Reaction. *Adv. Mater. Interfaces* **2020**, *7*, 2000396. [CrossRef]
10. Balint, L.-C.; Hulka, I.; Kellenberger, A. Pencil Graphite Electrodes Decorated with Platinum Nanoparticles as Efficient Electrocatalysts for Hydrogen Evolution Reaction. *Materials* **2021**, *15*, 73. [CrossRef]
11. Liu, M.; Wang, X.; Liu, J.; Wang, K.; Jin, S.; Tan, B. Palladium as a Superior Cocatalyst to Platinum for Hydrogen Evolution Using Covalent Triazine Frameworks as a Support. *ACS Appl. Mater. Interfaces* **2020**, *12*, 12774–12782. [CrossRef] [PubMed]
12. Chen, A.; Ostrom, C. Palladium-Based Nanomaterials: Synthesis and Electrochemical Applications. *Chem. Rev.* **2015**, *115*, 11999–12044. [CrossRef] [PubMed]
13. Yu, H.; Díaz, A.; Lu, X.; Sun, B.; Ding, Y.; Koyama, M.; He, J.; Zhou, X.; Oudriss, A.; Feaugas, X.; et al. Hydrogen Embrittlement as a Conspicuous Material Challenge–Comprehensive Review and Future Directions. *Chem. Rev.* **2024**, *124*, 6271–6392. [CrossRef]
14. Sathiyan, K.; Lal, A.; Borenstein, A. Laser Processed Magnetite Nanoparticles Embedded on rGO Composites for Efficient Electrocatalytic Oxygen Evolution Reaction. *Adv. Sustain. Syst.* **2022**, *6*, 2200076. [CrossRef]
15. Khine, Y.Y.; Wen, X.; Jin, X.; Foller, T.; Joshi, R. Functional groups in graphene oxide. *Phys. Chem. Chem. Phys.* **2022**, *24*, 26337–26355. [CrossRef]
16. Amirov, R.R.; Shayimova, J.; Nasirova, Z.; Solodov, A.; Dimiev, A.M. Analysis of competitive binding of several metal cations by graphene oxide reveals the quantity and spatial distribution of carboxyl groups on its surface. *Phys. Chem. Chem. Phys.* **2018**, *20*, 2320–2329. [CrossRef]
17. Amirov, R.R.; Shayimova, J.; Nasirova, Z.; Dimiev, A.M. Chemistry of graphene oxide. Reactions with transition metal cations. *Carbon* **2017**, *116*, 356–365. [CrossRef]
18. Chen, X.; Wu, G.; Chen, J.; Chen, X.; Xie, Z.; Wang, X. Synthesis of "Clean" and Well-Dispersive Pd Nanoparticles with Excellent Electrocatalytic Property on Graphene Oxide. *J. Am. Chem. Soc.* **2011**, *133*, 3693–3695. [CrossRef]
19. Hanan, A.; Shu, D.; Aftab, U.; Cao, D.; Laghari, A.J.; Solangi, M.Y.; Abro, M.I.; Nafady, A.; Vigolo, B.; Tahira, A.; et al. Co2FeO4@rGO composite: Towards trifunctional water splitting in alkaline media. *Int. J. Hydrogen Energy* **2022**, *47*, 33919–33937. [CrossRef]
20. Kumar, K.P.A.; Alduhaish, O.; Adil, S.F.; Pumera, M. Grafting of Pd on Covalently and Noncovalently Modified N-Doped Graphene for Electrocatalysis. *Adv. Mater. Interfaces* **2022**, *9*, 2102317. [CrossRef]
21. Zubir, N.A.; Yacou, C.; Motuzas, J.; Zhang, X.; da Costa, J.C.D. Structural and functional investigation of graphene oxide–Fe3O4 nanocomposites for the heterogeneous Fenton-like reaction. *Sci. Rep.* **2014**, *4*, 4594. [CrossRef] [PubMed]
22. Tang, S.; Zhu, J. Structural and electronic properties of Pd-decorated graphene oxides and their effects on the adsorption of nitrogen oxides: Insights from density functional calculations. *RSC Adv.* **2014**, *4*, 23084–23096. [CrossRef]
23. Karuppasamy, L.; Gurusamy, L.; Anandan, S.; Liu, C.-H.; Wu, J.J. Graphene nanosheets supported high-defective Pd nanocrystals as an efficient electrocatalyst for hydrogen evolution reaction. *Chem. Eng. J.* **2021**, *425*, 131526. [CrossRef]
24. Gao, T.T.C.; Wada, T.; Seenivasan, K.; Chammingkwan, P. *Critical Properties for Stabilizing Pd Nanoparticles in Graphene Oxide Framework-Supported Catalysts: Insights from Multifaceted Characterization*; Elsevier: Amsterdam, The Netherlands, 2024. [CrossRef]
25. Tran, T.P.N.; Thakur, A.; Trinh, D.X.; Dao, A.T.N.; Taniike, T. Design of Pd@Graphene oxide framework nanocatalyst with improved activity and recyclability in Suzuki-Miyaura cross-coupling reaction. *Appl. Catal. A Gen.* **2018**, *549*, 60–67. [CrossRef]
26. Karmakar, A.; Kundu, S. A concise perspective on the effect of interpreting the double layer capacitance data over the intrinsic evaluation parameters in oxygen evolution reaction. *Mater. Today Energy* **2023**, *33*, 101259. [CrossRef]
27. Mukherjee, P.; Sathiyan, K.; Bar-Ziv, R.; Zidki, T. Chemically Etched Prussian Blue Analog–WS_2 Composite as a Precatalyst for Enhanced Electrocatalytic Water Oxidation in Alkaline Media. *Inorg. Chem.* **2023**, *62*, 14484–14493. [CrossRef]
28. Schneider, C.A.; Rasband, W.S.; Eliceiri, K.W. NIH Image to ImageJ: 25 Years of image analysis. *Nat. Methods* **2012**, *9*, 671–675. [CrossRef]
29. Guo, S.; Liu, Y.; Murphy, E.; Ly, A.; Xu, M.; Matanovic, I.; Pan, X.; Atanassov, P. Robust palladium hydride catalyst for electrocatalytic formate formation with high CO tolerance. *Appl. Catal. B Environ.* **2022**, *316*, 121659. [CrossRef]
30. Rose, A.; Maniguet, S.; Mathew, R.J.; Slater, C.; Yao, J.; Russell, A.E. Hydride phase formation in carbon supported palladium nanoparticle electrodes investigated using in situ EXAFS and XRD. *Phys. Chem. Chem. Phys.* **2003**, *5*, 3220–3225. [CrossRef]

MDPI AG
Grosspeteranlage 5
4052 Basel
Switzerland
Tel.: +41 61 683 77 34

Materials Editorial Office
E-mail: materials@mdpi.com
www.mdpi.com/journal/materials

www.ingramcontent.com/pod-product-compliance
Lightning Source LLC
Chambersburg PA
CBHW041342210926
51850CB00013B/2954
* 9 7 8 3 7 2 5 8 3 8 8 7 5 *

Oil and Gas Well Engineering Measurement and Control